A COURSE IN
ELEMENTARY METEOROLOGY

Met. O. 911

METEOROLOGICAL OFFICE

A Course in Elementary Meteorology

SECOND EDITION

LONDON
HER MAJESTY'S STATIONERY OFFICE

U.D.C.: 551.5 (02)

© *Crown copyright 1978*

First edition 1962

Second edition 1978

HER MAJESTY'S STATIONERY OFFICE

Government Bookshops

13a Castle Street, Edinburgh EH2 3AR
49 High Holborn, London WC1V 6HB
41 The Hayes, Cardiff CF1 1JW
Brazennose Street, Manchester M60 8AS
Southey House, Wine Street, Bristol BS1 2BQ
258 Broad Street, Birmingham B1 2HE
80 Chichester Street, Belfast BT1 4JY

Government Publications are also available
through booksellers

£4.95 net

Printed in England for Her Majesty's Stationery Office
by McCorquodale Printers Ltd., London

HM 7908   Dd 586655 K40 5/78   McC 3309
ISBN 0 11 400312 2

PREFACE TO THE FIRST EDITION

IT is now forty years since the late W. H. Pick wrote the first edition of *A short course in elementary meteorology*. That book, describing the different classes of weather systems and their characteristics and behaviour, and explaining in simple language the physical processes which caused certain types of cloud formation, wind regimes and weather, was the forerunner of a number of 'popular' books on meteorology. Despite the competition, Pick's *A short course in elementary meteorology* never lost its appeal and sales have continued at a high level ever since its first appearance.

A short course in elementary meteorology has been revised from time to time but no major amendment has been made since the issue of the fifth and last edition in 1938. Since then, great advances have been made in meteorological knowledge, largely owing to the accumulation of observations at high levels in the atmosphere by means of aircraft and radiosondes and more recently to intensive research into the microphysics of clouds; and although meteorological conditions in the lowest layers are those which affect us most closely, the emphasis in a study of the physical processes of meteorology has shifted from the lower layers to higher levels. Some thought was given to the possibility of incorporating in a revised edition of *A short course in elementary meteorology* a chapter or two on the meteorology of the upper air, but the permeation of the science by modern concepts of conditions aloft necessitates an integrated approach to the subject as a whole. Accordingly it was decided to withdraw the book and write a completely new work to replace it.

The present book is the work of D. E. Pedgley, B.Sc., and was written while he was a member of the staff of instructors at the Meteorological Office Training School. It is intended for the reader whose knowledge of physics is roughly equivalent to that of upper science forms in schools, although parts of the book are of a rather higher standard. These are included, in smaller print, for the benefit of those who are able to assimilate them and others whose interest may be aroused and who may thereby be induced to study the subject more deeply. Mr Pedgley has had considerable experience in training staff who have entered the Meteorological Office with little or no previous knowledge of meteorology and in writing this book he has been able to draw on his experience. His work is confidently presented as an authoritative and methodical account of the elements of meteorology as they are taught today.

PREFACE TO THE SECOND EDITION

THIS second edition of *A course in elementary meteorology* is a revision by H. Heastie, M.Sc., written while he was an instructor on the staff of the Meteorological Office College. Some parts of the chapters on precipitation and forecasting have been extensively revised. Two appendices have been added to give a short introduction to the use of radar and satellites in meteorology at the present time.

As far as possible in this edition, SI (système international) units have been used; however, the use of certain non-SI units has been continued for specialized measurements (e.g. knot, nautical mile).

Grateful acknowledgement for permission to use material from their publications is made to the following:

Dr B. J. Mason, *Rain and rainmaking* (1st edition)—Figure 39,

National Hail Research Experiment Report 75/1, by K. A. Browning and G. B. Foote—Figure 45,

National Hail Research Experiment Report 76/1, by K. A. Browning and others—Figures 46–48.

CONTENTS

PART I

PHYSICAL METEOROLOGY

LIST OF DIAGRAMS

LIST OF PLATES

PART I
PHYSICAL METEOROLOGY

CHAPTER 1

TEMPERATURE

1.1 INTRODUCTION

1.1.1 *Measurement of temperature*

Of all the many elements of the weather which affect our daily lives, temperature is probably the most important. This is true not only from our direct contact with warm and cold air, but also indirectly because, as we shall see in later chapters, changes of temperature in the atmosphere largely control both the wind and the concentration of water vapour in the air, and these two in turn are all-important in determining the formation of clouds and rain.

Temperature is expressed by means of one of three scales, Fahrenheit (°F),* Celsius (°C) and kelvin (K). All three scales are used in meteorology and conversion from one to the other should be understood. If T represents the temperature of a body, then

$$T \text{ °F is equivalent to } \frac{5}{9}(T-32) \text{ °C}$$

and
$$T \text{ °C is equivalent to } (\frac{9}{5}T+32) \text{ °F.}$$

Also $\qquad T$ °C is equivalent to $(T+273)$ K

and $\qquad T$ K is equivalent to $(T-273)$ °C.

It follows from these relationships that 32 °F = 0 °C = 273 K and 212 °F = 100 °C = 373 K.

Temperature is measured with a *thermometer*; many types exist, the principles of those most commonly used in meteorology being described below.

(a) *Liquid-in-glass* type, where the differential expansion of a liquid with respect to its glass container is measured. Volume changes of the liquid are shown by changes in position of the end of a column of the liquid in a tube attached to the liquid's container. The liquid used depends upon the temperature range over which the instrument will be used; for ordinary purposes mercury is used but if the temperature is likely to fall below the melting-point of mercury (− 38·8 °C) then ethyl alcohol or some other organic liquid is employed.

(b) *Mercury-in-steel* type, where expansion of the mercury alters the shape of a hollow, flexible metal coil, which either unrolls or rolls up more tightly. The coil and bulb are connected by a thin, steel-cored capillary tube which may be more than 30 metres long, so enabling the instrument to be read at a considerable distance from the measuring point.

---

* Member states of the European Communities will cease to authorize the use of the Fahrenheit scale of temperature with effect from 31 December 1979 at the latest (EEC Directive of 27 July 1976).

(c) *Bimetallic-strip* type, where two pieces of metal having different coefficients of expansion are welded side by side so that a temperature change causes the strips to curl slightly, the extent increasing with the temperature.

(d) *Electrical resistance* type, depending upon the known variation of electrical resistance of a metal wire as the temperature changes.

(e) *Thermocouple* type, where two metals are joined in a closed electrical circuit, the two joints being at different temperatures, and the electro-motive force set up in the circuit is measured. This force varies with the temperature difference so that, if it is measured whilst one joint is kept at a standard temperature, the temperature of the other can be calculated.

A continuous and permanent record of temperature at a given place can be obtained by making the temperature-sensitive part of the instrument operate a pen-arm which marks a trace on a paper chart wrapped around a slowly rotating drum. Such an instrument is known as a *thermograph*, and its chart is a *thermogram*. Any instrument which gives a continuous and permanent record of the changes of some meteorological element is known as an *autographic* instrument. Other autographic instruments in widespread use are barographs (for recording atmospheric pressure), anemographs (wind speed and direction), hygrographs (humidity) and recorders for rate and duration of rainfall.

1.1.2 *Some definitions*

In meteorology we need to know the temperature both of the earth's surface and of its atmosphere. In particular, we shall be interested in temperatures of the air, the ground and the sea, each of which varies with place and time.

The term 'air temperature' is applied strictly to that of the air and may be measured by placing a thermometer in contact with the air. At the same time the thermometer must be protected from the effects of external heat sources. To compare readings from neighbouring places, each thermometer must be exposed to the air under identical conditions. This is best achieved by keeping them in carefully designed screens which give them standard surroundings but at the same time allowing the air ready access to them. The most important source of error is exposure to direct sunlight: a thermometer in the sun records a far higher temperature than one in the shade (see Section 1.2.1). The 'shade temperature' is usually a good approximation to the 'screen temperature'. A further point to bear in mind is that the thermometers should be kept at a standard height above the ground because on many occasions the temperature is found to vary markedly with height (see Section 1.3.2). In Britain, thermometers in the screen have their bulbs at 1·25 metres above the ground. Unless otherwise stated we shall use the term 'air temperature' to mean 'screen temperature'.

Frost is an important feature of our weather, the term being applied to several phenomena; it is important to distinguish between them.

(a) *Air frost*, a screen temperature below 0 °C.

(b) *Ground frost*, a temperature of 0 °C or below recorded by a thermometer lying horizontally with its bulb just in contact with the blade-tips of a turf surface, cut short.

(c) *White frost* or *hoar frost*, a deposit of white, feathery ice crystals (see Section 3.5.2).

(d) *Black frost*, a condition in which the temperature of the ground cools to below 0 °C with no deposit of hoar frost.

(e) *Glazed frost*, a layer of glassy ice (see Section 6.3.5).

(f) *Silver frost*, a deposit of frozen dew (see Section 3.5.2).

1.2 HEAT TRANSFER PROCESSES AND TEMPERATURE CHANGES

1.2.1 *Heat transfer processes*

When two bodies having different temperatures are placed near each other, heat flows between them and there is a tendency for the two temperatures to become equal. This transfer of heat may take place in three ways—by conduction, convection and radiation.

(a) *Conduction:* When the two bodies are in contact, some of the kinetic energy of the molecules in the hot body is transferred to the molecules in the cold body during collisions between them at their interface. The rate of flow of heat increases with the temperature difference, but it also depends upon the nature of the substance through which the heat is flowing. Air conducts heat slowly—it is a poor conductor. Rocks are relatively good conductors so we would expect soils, which are essentially mixtures of small particles of rocks with many air-spaces, to be only moderately good conductors. Water and ice are also moderate conductors so it follows that fresh lying snow, which contains a large proportion of air (about 10 parts of air to 1 of ice), is only a poor conductor. This explains how animals can survive fatally low air temperatures if they are buried in deep snowdrifts; similarly, the new growth on plants in spring is protected from frost by lying snow. A further illustration of the poor conductivity of air is shown by snow being able to settle and persist more easily on grass than on roads if the ground temperature is not below 0 °C. This is because the large volume of air among the grass blades prevents much contact between the snow and the ground.

(b) *Convection:* This involves a mass movement within a fluid (liquid or gas). It may be brought about in two ways: either by the fluid moving over a rough boundary surface and so inducing irregular eddies, known as *forced convection* or *turbulence*; or by the fluid being heated from below so that parts of it become less dense, and therefore buoyant, and they rise as upward flowing currents replenished by adjacent downward currents, the whole being known as *free convection*. Throughout this book, references to convection will be to this latter type. Turbulence is always present in the atmosphere wherever the wind is blowing, especially near the ground; convection occurs when the lowest layers of the atmosphere are in contact with warmer ground.

(c) *Radiation:* Heat can flow between two bodies, even when they are not in contact, in the form of electromagnetic radiation. This method does not require the presence of an intervening material medium but if one is present then it may alter the process.

1.2.2 *Some properties of radiation*

All bodies emit radiation continuously in the form of waves similar to radio waves but usually with much shorter wavelengths. Their wavelengths are so short that a new unit of length is used when describing them. This is the *micrometre* (symbol μm) equal to 10^{-6} metre. It is a useful unit, too, for measuring the sizes of the minute particles in clouds. Only a limited range of wavelengths can be detected by the eye in the form of light, namely those lying between the approximate limits of 0·4 μm and 0·7 μm, corresponding to violet and red light respectively. Radiation with wavelengths just less than 0·4 μm is known as ultraviolet radiation and that with wavelengths just above 0·7 μm is infrared; each is invisible.

Radiation emitted by a body at any given time covers a range of wavelengths. The intensity of radiation with a given wavelength varies with the wavelength, being greatest a little above the bottom of the range. Thus, 99 per cent of the sun's or *solar radiation* lies within the limits 0·15 μm and 4 μm, and has its maximum intensity at 0·6 μm, that is, within the visible range. Note that it also extends into both the infrared and the ultraviolet. Of the earth's or *terrestrial radiation*, 99 per cent lies within the approximate limits 4 μm and 100 μm with a maximum at about 15 μm, that is, it is wholly in the infrared and is therefore invisible. These differences in wavelength account for the commonly used terms 'short-wave' radiation for that from the sun and 'long-wave' radiation for that from the earth.

There are two laws referring to radiation which are useful to us:

(a) *Stefan's Law*, which states that the rate of loss of radiant heat by unit area of a body is directly proportional to the fourth power of its absolute temperature, so that hot bodies radiate very much more intensely than cold bodies.

(b) *Wien's Law*, which states that the wavelength of the radiation of maximum intensity is inversely proportional to its absolute temperature. This is illustrated by the colour changes of a poker as it gets hotter in the fire. First it is dark red, then orange as yellow wavelengths are mixed with the red, and finally white when all colours are mixed.

These laws account for the differing properties of solar and terrestrial radiation described above.

The intensity of the radiation emitted by the sun seems to be nearly constant. The solar radiation flux at a surface normal to the sun's beam outside the earth's atmosphere at the earth's mean distance from the sun is about 1350 W m^{-2} and is known as the *solar constant*. However, the intensity found at the ground is not constant because:

(a) some radiation is lost whilst passing through the atmosphere;

(b) the sun is rarely vertically overhead so the energy is spread over a ground area greater than 1 square metre; the smaller the elevation of the sun above the horizon the greater the area and, therefore, the weaker the intensity. The sun's elevation depends upon the time of day, the season and the latitude, so the intensity is relatively weak near sunrise or sunset, in winter and at high latitudes.

The losses incurred as solar radiation passes through the atmosphere are caused by:

(a) *Absorption* (about 15 per cent), especially by water vapour and ozone (see Section 1.3.2). The oxygen and nitrogen in the atmosphere may be considered as transparent.

(b) *Scattering* (about 10 per cent), or the alteration of the direction of the radiation as it passes very near the air molecules (see also Section 8.4.1).

(c) *Reflection* (about 30 per cent) from clouds, and from the ground especially when snow- or ice-covered.

The total effect is that only about 45 per cent of the radiation entering the atmosphere is absorbed by the ground. So, remembering that the cross-sectional area of the solar beam intercepted by the earth is one-quarter of the surface area of the earth, this implies that on average each square metre of the ground receives 150 watts.

Most instruments which measure the intensity of solar radiation depend upon the fact that the temperature of a body rises when it absorbs the radiation. They are called *radiometers* (*pyrheliometers* or *pyranometers*); several types are in use:

(a) *Silver-disc* type, where the rate of rise of temperature of a silver disc, exposed under standard conditions, increases with the intensity of the radiation.

(b) *Thermopile* type, where a series of thermocouples has one set of its junctions exposed whereas the other set is maintained at a constant temperature. The electromotive force produced increases with the intensity of the radiation.

(c) *Differential-absorption* type, where two different objects are exposed under identical conditions and the difference in their temperatures increases with the intensity of the radiation.

(d) *Bimetallic-strip* type, either where the change in shape of an exposed strip is measured under a microscope, or where two strips are exposed and, because one is black-coated whereas the other is white, they change shape to different extents.

An estimate of insolation intensity may be found by using a thermometer with a black-bulb-in-vacuo exposed to direct sunlight. The radiation is absorbed by the bulb, which is painted black to absorb the maximum amount of radiation and surrounded by a vacuum to minimize heat loss by conduction. When placed in the sun it records a high temperature, becoming steady when the rate of gain of short-wave radiation just equals the rate of loss of long-wave radiation. No two instruments read alike but a given instrument can give a rough guide to the intensity of insolation; the greater the intensity, the greater the black-bulb temperature. It may exceed 77 °C on a summer day.

1.2.3 *Heating the atmosphere*

All three methods of heat transfer are active in heating the atmosphere; they are shown diagrammatically in Figure 1. Radiation from the sun passes through the atmosphere with some depletion resulting from absorption, scattering and reflection, the remainder being absorbed by the ground which thus becomes hotter. As soon as it becomes hotter than the air above it, heat flows by conduction and the air becomes warmer. Because of its poor conductivity this heating of the air is confined to a very shallow layer near the ground, the layers above being

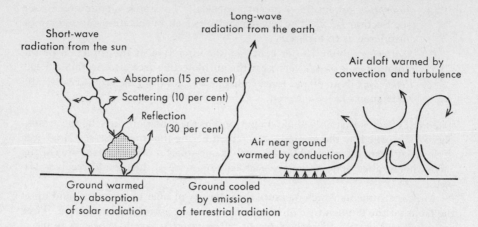

Short-wave radiation from the sun

Long-wave radiation from the earth

Air aloft warmed by convection and turbulence

Absorption (15 per cent)

Scattering (10 per cent)

Reflection (30 per cent)

Air near ground warmed by conduction

Ground warmed by absorption of solar radiation

Ground cooled by emission of terrestrial radiation

FIGURE 1. *Physical processes involved in heating the atmosphere*

heated by mixing with the shallow warm layer as a result of convection or turbulence.

It is most important to understand the part played by each of these processes. More detail will be given in Section 1.5.

1.2.4 *Specific heat*

When heat is added to a body by one of the methods described in Section 1.2.1, its temperature rises (unless a change of phase is induced in that body—see Section 3.1.2). The amount of heat required to raise the temperature of unit mass of a substance by one kelvin is known as the specific heat of the substance. It is normally measured in joules per kilogram per kelvin. The specific heat of water at 0°C is 4217 J kg^{-1} K^{-1} and of ice 2100 J kg^{-1} K^{-1}, and for soils it varies with the amount of water in the soil from about 1000 J kg^{-1} K^{-1} upwards. Using these values we see that a given quantity of heat supplied to 1 kilogram of ice will raise its temperature twice as much as if it were supplied to 1 kilogram of water; added to 1 kilogram of soil the temperature rise may be up to four times as great. This fact will be important when we come to consider temperature variations of the ground.

In general if Q joules of heat are added to m kilograms of a substance having a specific heat c J kg^{-1} K^{-1}, then the temperature rise θ is given by:

$$\theta = \frac{Q}{mc}.$$

1.2.5 *Adiabatic temperature changes*

There is another way by which the temperature of a body may be altered but this time no heat is added. This method is especially applicable to gases. When a mass of gas is compressed, and at the same time no heat is allowed to flow to or from its surroundings, then its temperature rises—the more it is compressed, the bigger the temperature rise. Conversely, if it is expanded the temperature falls. A change in the temperature of a mass of any substance without there being any transfer of heat between that substance and its surroundings is said to be an *adiabatic change.* Adiabatic changes of air in the atmosphere, brought about by changes of pressure, are extremely important because they produce most of our clouds and precipitation (see Section 1.4.2 and Chapters 5 and 6).

1.3 STRUCTURE OF THE ATMOSPHERE

1.3.1 *Composition of the atmosphere*

Numerous measurements of the composition of the atmosphere near the ground have been made, even as early as 1852 by Regnault. They show that the percentage composition by mass of clean 'dry air' (that is, air from which water vapour has been removed) is constant:

Constituent	Percentage (by mass)
Nitrogen	75·51
Oxygen	23·15
Argon	1·28
Other gases (including rare gases, ozone and carbon dioxide)	0·06
Total	100·00

Measurements at great heights, made either directly from balloons and rockets or indirectly by study of the wavelengths of radiation transmitted by the atmosphere, have shown that this composition changes very little with height. However, above about 80 kilometres some of the sun's ultraviolet radiation is absorbed by oxygen and nitrogen molecules which then decompose, either into free atoms or ionized molecules and atoms, forming a layer known as the *ionosphere* which extends upwards indefinitely. Its density is extremely small—the density ratio relative to the mean sea level value being about 10^{-5} at 80 kilometres and 10^{-10} at 200 kilometres.

Between about 20 kilometres and 50 kilometres more oxygen molecules are split but the resulting atomic oxygen recombines with unaffected molecules to give ozone. This layer containing ozone is known as the *ozonosphere*.

1.3.2 *Vertical temperature structure*

Regular measurements of temperature up to heights of 20 to 30 kilometres are now made twice daily over large areas of the earth's surface. These readings are obtained by using a *radiosonde*, the British form of which consists of a sounding balloon and attached to it is a thermometer and a small radio-transmitter which telemeters information back to a ground receiving station in the form of an audio-frequency note, whose frequency changes with the temperature. Pressure and humidity are also measured by the radiosonde. Eventually the balloon bursts and the instruments descend by parachute. Their recovery is not essential to an understanding of their readings as it was with the older methods of measuring temperature high up in the atmosphere. Then, either kites or balloons were used carrying autographic instruments. A few flights in manned balloons were also made. The radiosonde, first developed during the 'thirties, is still supplemented to a small extent by aircraft observations (see also Appendix B, Section B.2.3).

If a graph is prepared showing the distribution of temperature with height on a given occasion, we find that, for different layers in the atmosphere, this distribution takes one of three forms:

(a) A *temperature lapse*—a decrease of temperature with height; the *lapse rate* is the rate at which temperature decreases with height, expressed in °C per kilometre.

(b) A *temperature inversion* (commonly just referred to as an 'inversion')—an increase of temperature with height; it has a negative lapse rate.

(c) An *isothermal layer*—the temperature does not vary with height; it has a zero lapse rate.

A typical distribution of temperature with height, as found by a radiosonde, is shown in Figure 2. We can immediately distinguish two layers in the atmosphere

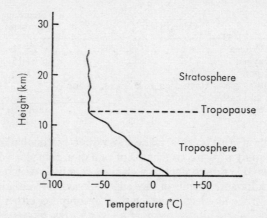

FIGURE 2. *Example of a temperature sounding through the lower atmosphere*

characterized by differences of lapse rate. Below about 11 kilometres there is a positive lapse rate on the average and this layer is known as the *troposphere*. At about 11 kilometres there is a well-marked boundary known as the tropopause. Above it lies the stratosphere, in which the temperature is nearly constant up to about 20 kilometres and then increases with height. Nearly all temperature soundings show the following features:

(a) *Troposphere.* Positive lapse rate of about 6 to 8 °C per kilometre. There may be one or more shallow layers, up to about 1 kilometre deep, which may be either isothermals or inversions.

(b) *Lower stratosphere.* Isothermal, or only a small lapse rate, either positive or negative.

(c) *Tropopause.* The boundary surface between the two layers. It is not horizontal, being lower over the poles than over the equator—approximately 6 to 8 kilometres and 16 to 18 kilometres respectively. Its height also varies with the season, in middle latitudes especially, being lower in winter than in summer—approximately 8 to 10 kilometres and 12 to 14 kilometres, respectively, over Britain. When the tropopause is low the stratosphere is relatively warm, −40 to −60 °C; when high it is cold, −60 to −80 °C.

The troposphere contains nearly all the 'weather' as we understand it, namely clouds, precipitation and winds, particularly the vertical winds.

Many people have investigated the problem of why the lower atmosphere exists in these two layers. The earlier ideas of Gold in 1909, Emden in 1913, and others sought to explain the stratosphere and troposphere as being the result of radiative and convective equilibrium respectively, the former involving both solar and terrestrial radiation and their absorption by water vapour in the air, whilst the troposphere was thought to result from convection through contact with a warm ground. In 1946, Dobson, Brewer and Cwilong pointed out the importance of ozone as an absorber and emitter of radiation but Goody, in 1949, suggested that water vapour and carbon dioxide are more important, the first constituent tending to cause a heating of the atmosphere and the second a cooling. It cannot be said that a full explanation has yet been obtained.

1.3.3 *The upper atmosphere*

Above about 30 kilometres other methods must be used to investigate the distribution of temperature with height. These have given somewhat divergent results but not all the differences may be real. However, a generalized picture is

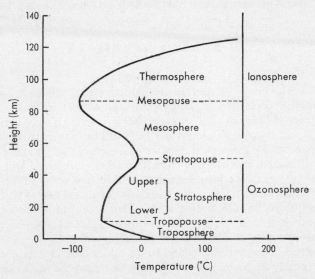

FIGURE 3. *Approximate distribution of temperature with height in the atmosphere*

shown in Figure 3. The methods used to measure or estimate these temperatures include:

(a) Rockets, where pressure and density readings are telemetered to the ground and the temperatures are deduced from them. Some of these rockets have been launched from high-altitude balloons.

(b) Observation of anomalous propagation of sound, in which the refraction, by different layers in the atmosphere, of sound waves from a large explosion, results in rings of audibility at ground level around the source, with intervening rings of silence. Murgatroyd in Britain in 1956 and Crary in the United States of America in 1950 are among those who have used this method.

(c) Releasing grenades at intervals from an ascending rocket. Both methods (b) and (c) rely upon the dependence of the speed of sound upon the density of air.

(d) Scattering of light from a vertically pointing searchlight at night, the extent depending upon the density of the air.

(e) Visual and radar observations of meteor trails.

(f) Radio propagation through the ionosphere.

(g) Absorption measurements from satellites (see Appendix B, Section B.2.3). It is hoped to extend the height range up to 1 millibar (50 kilometres) by 1978.

The existence of the two warm layers, the ionosphere and the ozonosphere, arises from the absorption of ultraviolet radiation from the sun.

In 1958, both Chapman and Nicolet, using data from artificial earth satellites, first suggested that our atmosphere may be continuous with the sun's, with its temperature almost constant with height above 400 kilometres at about 1200 °C, rising slowly towards the sun. Continuity of the two atmospheres allows direct conduction of heat between the sun and the earth but it is so slow as to be negligible.

1.4 STATIC STABILITY

1.4.1 *Static stability*

Study of the processes involved in the formation of clouds leads us to consider a fundamental concept in meteorology known as *static stability*. It refers to the problem of what happens to a parcel of air after it has been given an initial vertical displacement, either upwards or downwards. If, after an upward displacement, the parcel is found to be warmer (and therefore less dense) than its surroundings, its buoyancy will make it move farther from its original position; it will continue to move upwards until its temperature somehow becomes equal to that of its surroundings. When, after an initial displacement, the parcel moves farther from its original level, then the surrounding atmosphere is said to be *statically unstable* (or commonly just 'unstable'). An initial disturbance in such an atmosphere results in spontaneous development of the disturbance. On the other hand, if, after displacement, the parcel is found to be colder (and therefore more dense) han its surroundings, its buoyancy will tend to return the parcel to its original level and then the surrounding atmosphere is said to be *statically stable* (or 'stable'). A disturbance in such an atmosphere must in the first instance be forced, since buoyancy tends to oppose any vertical motion. There is clearly an intermediate state where vertical motion is neither encouraged nor opposed; the atmosphere is then said to be *neutrally stable*.

These ideas on static stability have been based on an upward-moving parcel of air; it is left to the reader to show that they are also true for a downward-moving parcel.

We may compare the parcel's motion to that of a ball-bearing in a watch-glass. If the glass is concave upwards, the ball is in equilibrium only at the bottom and any sideways displacement results in a return to the centre. If the glass is convex upwards, the ball remains stationary when it is at the centre, but any sideways displacement results in further movement away from the centre. In the first state the ball is in a stable environment; in the second, unstable. If the glass were flat there would be neutral stability.

1.4.2 *Adiabatic lapse rates*

A parcel of air rising through the atmosphere expands because the pressure exerted upon it decreases with height. This expansion is approximately adiabatic for two reasons: air is a poor conductor of heat, and mixing of the parcel into its surroundings is slow. Adiabatic expansion results in cooling, so the parcel cools as it ascends. Calculations show that the rate of cooling of a parcel of air as it expands adiabatically during ascent is 10 °C per kilometre as long as no water vapour in the air condenses as the parcel ascends. This lapse rate is known as the *dry adiabatic lapse rate* (DALR) and is constant no matter what the original temperature of the parcel may be. A descending parcel warms at the same rate as a result of adiabatic compression.

Consider now the surroundings of a rising parcel. If they are cooler than the parcel at all levels (other than the original, where the two temperatures are the same) then it is unstable because the parcel is always warmer, after its initial displacement. For this to be true the *environment lapse rate* (ELR) must be greater than the lapse rate of the parcel. So, in order to determine whether the atmosphere is unstable or not, all we need to know are the lapse rates of the parcel and of its environment. Remember, the parcel's lapse rate is a consequence of its ascent; the ELR arises from other factors as yet unconsidered. We see then that if the ELR is greater than the DALR, the surroundings are unstable for ascent or descent of dry air, that is, of air which does not contain liquid water or ice suspended in it. Also, if the ELR is less than the DALR, the surroundings are stable.

When the parcel ascends so far that the cooling causes condensation of some of the water vapour within it, then the cooling is partly offset by the release of latent heat (see Section 3.1.2). Thus, air rising adiabatically, with at the same time some of its water vapour condensing within it, cools more slowly than dry air. The rate at which it cools, known as the *saturated adiabatic lapse rate* (SALR), is about 6 °C per kilometre at ordinary temperatures but it is not a constant. It has a small value at high temperatures but becomes nearly equal to the DALR below −40 °C. This is because at these low temperatures the maximum amount of water vapour which can be present in the air is very small (see Section 3.1.4) so that, for a given fall in temperature, the amount of water vapour which condenses (and hence the amount of latent heat liberated) is so small that it hardly alters the cooling resulting from expansion. Note that the parcel will warm at the same rate when it is compressed adiabatically by descent only as long as sufficient liquid water or ice is present to evaporate and keep the air saturated.

Consider now the surroundings of this parcel whose water vapour condenses on ascent. Arguing as above, if the ELR is greater than the SALR, the surroundings are unstable for ascent or descent of saturated air; and if the ELR is less than the SALR, the surroundings are stable.

It is clear, then, that the stability of the environment depends not only upon its lapse rate but also upon whether the rising parcel is saturated or not.

An interesting state arises when the ELR is less than the DALR but greater than the SALR, for the surroundings are then stable as long as the parcel is unsaturated but unstable if it is saturated. Such a state is rather common; when it occurs the atmosphere is said to be *conditionally unstable*.

Isothermal layers and inversions are extremely stable whether the rising parcel is dry or not.

1.4.3 *Development of instability and stability*

The ELR of the atmosphere is not constant—it varies with both space and time. On some occasions the atmosphere is unstable and on others it is stable. Since the idea of stability is so important in meteorology it will be useful to consider the processes which can change the ELR.

The ELR of a layer in the atmosphere increases when either the lower part is warmed or the upper part is cooled—or both occur together. Warming of the lower part may be the result of contact with a warm, underlying ground or of the replacement of the original air by new, warmer air brought in by the wind. Cooling of the upper part may be the result of heat loss by long-wave radiation or of its replacement by new, cooler air. Direct loss of heat by long-wave radiation is slow, except when clouds are present (see, for example, Section 5.3.3).

Conversely, the ELR of a layer decreases when either the lower part is cooled or the upper part is warmed, or both. Cooling of the lower part may be the result of contact with a cold, underlying ground or of its replacement by new, cooler air. Warming of the upper part may be the result of absorption of short-wave radiation or of its replacement by new, warmer air. Direct warming by short-wave radiation is slow, even when clouds are present.

Many examples of stability changes will be discussed on later pages.

1.5 DIURNAL TEMPERATURE VARIATIONS

1.5.1 *The normal variation*

In Section 1.2.3 we saw that the atmosphere is heated by the sun but to a large extent only indirectly, the solar radiation first warming the ground. Also, it is cooled by contact with the ground which itself has been cooled by loss of terrestrial radiation. Direct absorption of insolation by the atmosphere is not important below the tropopause and calculations show that a warming of only about 0·5 °C per day is possible. Direct cooling by loss of long-wave radiation may amount to 1 or 2 °C per day. These changes are small compared with those produced near the ground by conduction, convection and turbulence.

The importance of the ground in the heating of the atmosphere leads us to consider it first. The *diurnal variation* of ground temperature, its variation throughout the 24 hours, follows a well-known pattern. Figure 4 gives an example for a cloudless and windless day, where it is seen that there is a minimum just after dawn and a maximum in the early afternoon. A curve showing the diurnal variation of *air* temperature near the ground is also given. It is similar in shape but is both a little out of phase (the minimum and maximum occur a little later) and smaller in amplitude (the minimum has a higher value, and the maximum a lower one).

The air temperature falls whilst the ground is colder than the air; it rises when the ground is warmer. Hence, the ground is colder than the air above it until the time of minimum air temperature and it is warmer from then until the time of maximum air temperature. One may easily verify these facts by making sets of simultaneous readings of ground and screen temperatures.

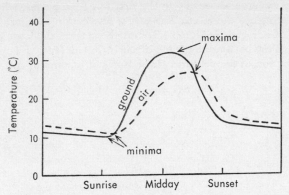

FIGURE 4. *Diurnal temperature variation of bare ground and air near the ground on a cloudless and windless day*

Sunrise and sunset are taken as 0600 and 1800 local time.

Consider now the shapes of these curves. From midnight to sunrise the ground loses heat by long-wave radiation but at the same time it gains heat by conduction both from the warmer air above and from deeper layers in the ground. However, these last two processes are slow and may be ignored as a first approximation, so the ground loses heat continuously and its temperature falls at a rate of perhaps 1 °C per hour. After sunrise, the ground has another source of heat—insolation—so it cools more slowly until about a half-hour after sunrise when the rate of loss of heat by long-wave radiation just balances the rate of gain by short-wave radiation and the fall in temperature is checked. Thereafter, the intensity of incoming solar radiation exceeds the intensity of the outgoing terrestrial radiation and so the ground becomes warmer: a minimum thus occurs about a half-hour after sunrise.

The ground temperature soon rises above that of the air (within about a further quarter-hour at which time the air has its minimum temperature) and then increases rapidly, until about 1300 to 1400 local time when insolation is again balanced, but this time conduction (to the air above and to the now relatively cooler, deeper layers of the ground) adds significantly to the terrestrial radiation. Because there is a balance a maximum occurs at this time. Between then and sunset the intensity of incoming radiation decreases so quickly that it is incapable of balancing the other processes, so the ground temperature falls quickly. At about 1400 to 1500 local time it has decreased to a value equal to that of the air temperature which now reaches its maximum. After sunset there is no solar radiation but, since heat losses by conduction are now small again, the ground temperature falls more slowly by loss of long-wave radiation.

Notice that whereas solar radiation is received only during daylight hours, terrestrial radiation is lost throughout the 24 hours.

The minimum ground temperature is usually only a few degrees Celsius below the minimum air temperature but the maximum ground temperature may be 30 °C or even more above the maximum air temperature. Air just in contact with the ground has a temperature nearly the same as that of the ground so that at night there is a well-marked inversion from the ground upwards, usually to several hundred feet, but during the day the lapse rate from the ground to screen level may be very large, especially in the first few centimetres, where it may be several hundred times the DALR. These first few centimetres are important in the formation of mirages (Section 8.4.2), but above them the lapse rate is usually just greater than the DALR up to the top of the turbulent layer. Figure 5 shows the vertical distribution of temperature at the times of maximum and minimum air temperatures.

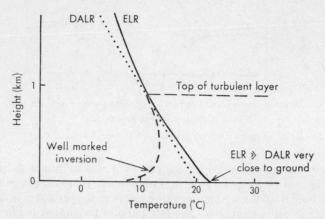

FIGURE 5. *Example of the vertical distribution of air temperature near the ground at times of maximum and minimum temperature at screen level*

In the turbulent layer the full and dashed lines represent the ELR at time of screen maximum and minimum respectively

1.5.2. *Controlling factors*

The amplitude of the diurnal variation is governed by many factors, some of which are noted here.

(a) *Amount of insolation*—dependent on both intensity and duration. As we saw in Section 1.2.2, the intensity will generally be less than the maximum possible. Absorption and scattering do not vary much, but reflection from clouds is variable and is very important; thick clouds allow only a small fraction of the insolation to pass through them. Even with clear skies, the intensity is relatively small in winter. The same is true of high latitudes at all seasons, but the long duration of summer sunshine there offsets the effect of low intensity.

(b) *Intensity of terrestrial radiation.* Water vapour is a good absorber of long-wave radiation and it re-radiates some of its absorbed heat back to the ground. A dry part of the atmosphere thus allows more radiation to escape into space than does a part containing much water vapour. Carbon dioxide acts similarly. We may be thankful that these two gases are present in the air for their absence would allow an increased rate of loss of long-wave radiation and the resulting cooling would be excessive, leading to very low night minima. Water vapour does not absorb much solar radiation so its total effect may be compared with the glass of a greenhouse—letting in short-wave but partially stopping long-wave radiation. This 'greenhouse effect' of the water vapour maintains our temperatures higher than they would be in its absence. Clouds are even better absorbers. Thick clouds re-radiate downwards a large part of the long-wave radiation absorbed by them (at wavelengths corresponding to their own temperatures) so that clouds are very effective in lessening the diurnal variation of temperature. With clear skies, the long nights of winter allow a large loss of heat by long-wave radiation.

(c) *Nature of the surface.* A land surface warms and cools far more quickly than a water surface, even though the rate of gain or loss of heat is the same. There are three reasons for this: firstly, the land has a lower specific heat; secondly, it is a much poorer conductor of heat; and thirdly, insolation raises the temperature of the top few centimetres of soil only, whereas the transparency of water allows heating down to greater depths. Night-time cooling is also confined to the top few centimetres of soil but as the top surface of water cools it becomes more dense, sinks and is replaced by warmer water from below. The diurnal variation of sea temperature is so small as to be difficult to measure: it is only 1 °C or less. Desert soils warm and cool the most quickly because they contain so much air, a poor conductor of heat. Moist soils change their temperatures slowly and soils covered with vegetation even more slowly. It is interesting to note that it is the uppermost parts of vegetation which are the effective absorbers and emitters of radiation, not the ground itself. The highest temperatures by day and the lowest by night are found near the vegetation's top. At night, the minimum is lower over vegetation than over bare ground because the air trapped by the stems and leaves acts as a very poor conductor of heat upwards from the relatively warm soil. Snow- or ice-covered ground is a good reflector of insolation but also it is a poor conductor, so the diurnal variation is large.

(d) *Depth of atmospheric layer through which heating is spread.* This increases with both the wind speed and the instability of the layer. When a deep layer is warmed the temperature rise is much smaller than when a shallow layer is warmed by the same amount of heat. A smaller temperature rise occurs, therefore, either when the wind is strong or when the atmosphere near the ground is very unstable. Cooling is influenced similarly. The shallow inversion present near the ground after a clear, calm night is destroyed by heating during the following morning when the air temperature may be observed to rise rapidly. After destruction of the inversion, heat is distributed through a deeper layer so its temperature rises more slowly.

As a summary, we can see that the following favour a large diurnal variation— no cloud, summer season, low latitudes, dry air, dry soil (or a snow surface), no wind and a stable atmosphere; and the following favour a small variation— thick cloud, winter season, high latitudes, moist air, wet soil (or better, a water surface), strong winds and an unstable atmosphere. Since many of these conditions vary markedly over Britain, it is no wonder that our temperature changes are diverse.

One last point: it is important to realize that we have been considering local temperature changes only; changes may also be brought about by *advection*, the flowing-in of air already having a different temperature, from somewhere else.

This is especially important near coasts and during the passage of a front (see Sections 1.6.1 and 9.6).

1.6 SPATIAL DISTRIBUTIONS OF AIR TEMPERATURE

1.6.1 *Air masses and fronts*

A good idea of the horizontal distribution of air temperature about a place may be obtained by plotting on a map values of temperature measured simultaneously and, say, at screen level. By drawing *isotherms*, or lines joining places with the same temperature, the cold and warm areas are immediately apparent and so are those places where the *temperature gradient* is greatest. The temperature gradient is the rate at which temperature varies in a horizontal direction perpendicular to the isotherms; where the isotherms are close, the gradient is large.

If we consider a chart covering a large area, say a square of side 5000 kilometres, then we can usually find two or more *air masses*, each covering many millions of square kilometres, and each characterized by its own temperature. This temperature is never exactly uniform over the whole of an air mass but it lies within a comparatively short range and the temperature gradients are small, perhaps 1 °C per 100 kilometres. At the boundaries of these air masses the gradients are much larger, perhaps 1 °C per 10 kilometres. These boundaries, where the contrast between neighbouring air masses is most concentrated, are known as *frontal zones*. They are narrow compared to the extent of the air masses themselves, say 100 kilometres compared with 1000 kilometres. Frontal zones will be considered in much more detail in Part II of this book but here it is useful to know that they slope upwards from the ground in such a way that the colder air mass underlies the warmer in the form of a very shallow wedge, whose angle is often less than one degree. Figure 6 represents a vertical cross-section through a frontal zone. A frontal zone is often associated with vertical motion in the atmosphere, particularly upward motion in the warm air mass, and it may then be accompanied by extensive cloud and precipitation. Figure 6 shows that the isotherms within both air masses slope gently from left to right; therefore, the air masses have small horizontal temperature gradients within

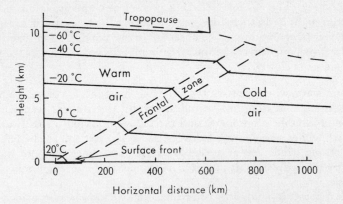

FIGURE 6. *Vertical cross-section through a frontal zone*
Note the different horizontal and vertical scales.

them, also from left to right. In the frontal zone they slope more quickly so that the gradients are greater there. Where the frontal zone intersects the ground we have a *front*.

1.6.2 *Local variations*

Air temperature is often found to vary markedly from place to place but on a much smaller scale than one of air masses and fronts. This is brought about by local variations in such factors as the nature of the ground, exposure to wind, and topography. Examples of such differences found in Britain are described in the *Meteorological Magazine* (**82**, 1953, p. 185 and p. 270, and **85**, 1956, p. 79) and in *Weather* (**8**, 1953, p. 69, and **9**, 1954, p. 82).

BIBLIOGRAPHY

ARMSTRONG, J.J.; 1974. Temperature differences between two ground-level sites and a roof site in Southampton. *Met Mag*, **103**, pp. 360–368.

DAVIES, J.W.; 1975. The relationship between minimum temperatures over different ground surfaces. *Met Mag*, **104**, pp. 78–85.

EBDON, R.A.; 1971. Periodic fluctuations in equatorial stratospheric temperatures and winds. *Met Mag*. **100**, pp. 84–90.

FRITH, R.; 1968. The earth's higher atmosphere. *Weather*, **23**, pp. 142–155.

GLOYNE, R.W.; 1971. A note on the average annual mean of daily earth temperature in the United Kingdom. *Met Mag*, **100**, pp. 1–6.

HARRISON, A.A.; 1971. A discussion of the temperatures of inland Kent with particular reference to night minima in the lowlands. *Met Mag*, **100**, pp. 97–111.

MANLEY, G.; 1974. Central England temperatures: monthly means 1659 to 1973. *Q J R Met Soc*, **100**, pp. 389–405.

MELVYN HOWE, G.; 1955. Soil-warmth in sunny and shaded situations. *Weather*, **10**, pp. 49–52.

MOCHLINSKI, K.; 1970. Soil temperatures in the United Kingdom. *Weather*, **25**, pp. 192–200.

MOORE, J.G.; 1956. Average pressure and temperature of the tropopause. *Met Mag*, **85**, pp. 362–368.

MURGATROYD, R.J.; 1970. The physics and dynamics of the stratosphere and mesosphere. *Rep Prog Phys*, **33**, pp. 817–880.

RIDER, N.E.; 1957. A note on the physics of soil temperature. *Weather*, **12**, pp. 241–246.

SHAW, J.B.; 1955. Vertical temperature gradient in the first 2,000 ft. *Met Mag*, **84**, pp. 233–241.

SMITH, K.; 1970. The effect of a small upland plantation on air and soil temperatures. *Met Mag*, **99**, pp. 45–49.

ZOBEL, R.F.; 1966. Temperature and humidity changes in the lowest few thousand feet of atmosphere during a fine sunny day in Southern England. *Q J R Met Soc*, **92**, pp. 196–209.

CHAPTER 2

PRESSURE AND WIND

2.1 ATMOSPHERIC PRESSURE

2.1.1 *Some definitions*

By the term *pressure* is meant the force exerted on unit area of a given surface. Its unit of measurement is, therefore, the unit of force divided by the unit of area, that is the newton per square metre ($N\ m^{-2}$), which has been given the name *pascal* (Pa). In meteorology the *millibar* (mb) is usually used as the unit of pressure; 1 mb = 100 Pa. Atmospheric pressure is not constant; the mean value over Britain at sea level is about 1013 millibars.

Now we have seen that the atmosphere comprises a mixture of gases, the composition, when all water vapour and impurities have been removed, being constant. Each of the constituents contributes to the total pressure of the mixture, that is, each has a *partial pressure*. Thus the partial pressure of the nitrogen is about 750 millibars, of the oxygen about 230 millibars and of the water vapour about 10 millibars (but this last is very variable—see Section 3.1.4). The partial pressure of a constituent in a mixture is the same as the pressure which that constituent would exert if it were present alone and still occupying the same volume (this is Dalton's Law), so that we can speak of the pressure of the water vapour, for example, without needing to take account of any effects from other constituents.

Since the pressure at any place in the atmosphere is the result of the weight of air above that place, pressure must decrease with height because as one ascends there is less air above. Figure 7 shows the approximate relation between pressure and height up to 20 kilometres, and it will be seen that the rate of decrease with

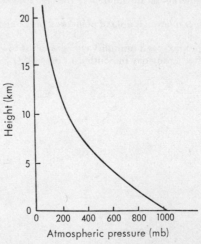

FIGURE 7. *Variation of atmospheric pressure with height*

20

height is not constant. Near the ground it is approximately 1 millibar per 10 metres but at 20 kilometres pressure decreases more slowly at about 1 millibar per 130 metres. At the top of a 1000-metre mountain, pressure is about 900 millibars.

The exact relation between pressure and height varies with place and time but can be computed from a knowledge of how pressure varies with temperature as measured by a radiosonde.

2.1.2 *Measurement of atmospheric pressure*

First, we will determine the pressure at the bottom of a column of fluid. Consider a column of height h metres and cross-section a square metres. Let it be filled with a fluid of density ρ kg m^{-3}. The pressure, p pascals exerted by the fluid on the base of the column is given by

$$p = \frac{\text{weight of fluid}}{\text{area of cross-section}}.$$

Taking g as the gravitational acceleration (approximately 9·81 m s^{-1}), the weight of the fluid is g times its mass, that is,

$$\text{weight of fluid} = g \times (\rho \times \text{volume}) \text{ newtons}$$
$$= \rho gah \text{ newtons.}$$

Hence
$$p = \frac{\rho gah}{a} \text{ pascals}$$

that is,
$$p = \rho gh \text{ pascals.} \qquad \ldots \ldots (1)$$

Now liquids are almost incompressible so that ρ is almost the same at each level in the column, but gases are compressible so that ρ itself decreases with height. We may still use equation (1) with a gas, however, if we take ρ to represent a mean density.

In 1643, Torricelli experimented with a column of mercury in a glass tube, closed at one end, and with its lower, open end dipped below the surface of a cistern full with mercury. He found that the length of the mercury column was nearly constant and independent of the length of the vacuum space occupying the upper part of the tube. The column is supported by the atmosphere, whose pressure, exerted downwards on the surface of the mercury in the cistern, just balances the pressure of the column. An increased atmospheric pressure can support a larger column.

A rough value for the length of this column can be calculated as follows. Knowing that the density of mercury is about 1·36 × 10^4 kg m^{-3}, g is about 9·81 m s^{-2} and atmospheric pressure is about 10^5 pascals, it follows from equation (1) (since p just balances atmospheric pressure) that

$$h = \frac{10^5}{1 \cdot 36 \times 10^4 \times 9 \cdot 81} \text{ m, that is approximately 0.75 m.}$$

Reversing the argument, we may use equation (1) to calculate the atmospheric pressure from a measurement of the height of the column. It may be expressed in 'centimetres (or inches) of mercury' and read from a scale engraved on the tube; however, this scale is best graduated directly in millibars.

An instrument which measures atmospheric pressure is known as a *barometer* and this particular type, employing a column of mercury, is a *mercury barometer*. There are many forms; the one used by the Meteorological Office is the Kew-pattern barometer (see the *Observers' Handbook*, 1969, p. 96, and the *Handbook of meteorological instruments*, Part I, 1956, p. 21). It is very important to notice that the barometer is made to read correctly only when ρ and g have certain fixed values. If, when the barometer is used, the prevailing values of ρ and g are different from those for which the instrument was calibrated, then the barometer reading must be corrected to account for the differences. Such corrections are essential if the pressure is to be measured to the accuracy required, namely 0·1 millibar, which is one part in 10 000. A further source of error in any sensitive instrument is the so-called 'index error', resulting from small inaccuracies during manufacture. It varies along the length of the scale and may be determined by comparing readings of the barometer with those of a very accurate standard instrument kept at the National Physical Laboratory.

Thus, to obtain a true value of atmospheric pressure at cistern level, we need to know not only the barometer reading but also corrections to account for:

(a) Index error.

(b) The current value of the density of mercury. This decreases as its temperature increases, so the atmosphere can support a longer column of warm mercury than cold; that is a barometer warmer than the calibration temperature reads too high.

(c) The local value of g. Since the earth is not perfectly spherical but rather flattened at the poles and also because it is rotating, the gravitational acceleration decreases from pole to equator, approximate values being 9·83 m s^{-2} at the poles and 9·78 m s^{-2} at the equator. It also decreases as height above the ground increases. For instruments used near sea level the former effect is the more important.

Variations of pressure with distance along the horizontal are much smaller than those in the vertical, being in the order of 1 millibar per 100 kilometres compared with 1 millibar per 10 metres. Although so small, horizontal variations are of fundamental importance in controlling the strength and direction of the horizontal wind, as we shall see in Section 2.2. They are best obtained by comparing readings from barometers kept in the same horizontal level. In practice, each barometer is at a different level, so to compare readings each must be corrected to some standard horizontal reference level, usually mean sea level. A barometer above mean sea level will thus have a correction added to its reading equal to the pressure of the imaginary column of air extending from the cistern down to mean sea level. From equation (1) it follows that this correction depends upon both the height of the barometer cistern above mean sea level and the density of the air imagined to fill the column; this latter, in turn, depends upon the total pressure and upon the temperature of the imaginary column (taken as screen-level temperature).

For any given barometer and place the index error, the height of the cistern and the local value of g are fixed so that, to obtain the pressure corrected to mean sea level, all we need are:

(a) the barometer reading,

(b) the barometer temperature,

(c) the screen-level temperature.

Combined corrections are worked out in advance for all likely values of these quantities and the results are made into a simple 'correction card'. Note that the changing of either barometer or place means that a new card must be used.

The *aneroid barometer* is another instrument for measuring atmospheric pressure. In principle, it consists of an evacuated, thin-walled, metal box which expands slightly as the pressure falls and contracts when the pressure rises; a powerful spring prevents the box from collapsing completely. Small changes in the shape of the box are magnified by a lever system operating a pointer which moves over a scale. Its main advantage over a mercury barometer is its portability, making it suitable for radiosondes, aircraft and expeditions.

The precision aneroid barometer (see The *Observer's Handbook*, 1969, p. 101) has largely superseded the mercury barometer for everyday use in the Meteorological Office. This instrument is simply a very sensitive form of the ordinary aneroid barometer. The capsule deflects a pivot bar which exerts a very much smaller resistance than that imposed by a conventional system of gears and levers. The movement of the bar is measured by a micrometer screw graduated in millibars. Contact between the bar and the micrometer is sensed electrically and displayed by means of a cathode-ray indicator tube. Readings from this type of barometer do not require correcting for gravity or for temperature.

An *autographic* barometer is a *barograph*; its pressure-sensitive element is usually an aneroid capsule.

2.1.3 *Isobars*

The horizontal distribution of mean-sea-level pressure may be represented on a map by drawing lines through places which have the same value of pressure. These lines are known as *isobars*; they represent the pressure distribution just as contours on an ordinary surface map represent the relief of a land surface. On a meteorological map isobars of mean-sea-level pressure are normally drawn. In mountainous country the errors introduced when calculating the correction to mean sea level for a high-level station are so large that cistern-level pressures are usually corrected to some other reference level, for example 1000 metres. In this book we will take all isobars as referring to mean-sea-level pressure unless stated otherwise.

When simultaneous values of pressure, corrected to mean sea level, obtained from barometers distributed over a large area are plotted on a map, it is often found that some of the isobars drawn to fit the values form closed curves around centres of either high or low pressure. A centre of high pressure is known as an *anticyclone* or *high*, whilst a centre of low pressure is a *depression* or *low*. The term 'cyclone' now has a restricted meaning and should not be used unless in that sense—see Section 10.4.2. An outward extension in the isobars from a centre is known as a *ridge* if from a high and as a *trough* if from a low. The area of almost constant pressure (and therefore few isobars) between two highs and two lows is known as a *col*. Isobar patterns illustrating these features are shown in Figure 8.

Most highs and lows each have several closed isobars when they are drawn at 2-millibar intervals. Even so, the absolute differences in pressure between the centres of neighbouring highs and lows are small. Central pressures of lows below 940 millibars or highs above 1050 millibars are exceptional. Over Britain the lowest pressure recorded was 925.5 millibars (corrected) at Ochtertyre, near

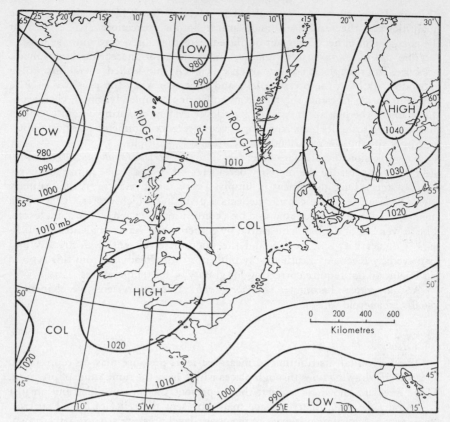

FIGURE 8. *Some common patterns of mean-sea-level isobars*

Crieff, Perthshire, on 26 January 1884; the highest was 1054·7 millibars at
Aberdeen on 31 January 1902.

Much more will be said about pressure systems in Part II of this book.

2.1.4 *Tendency*

Measurements of atmospheric pressure made regularly at a given place for
a long period (say, hourly for several days) show that it is seldom constant for
long. The rate of change of pressure with time is known as the *barometric tendency*.
It is not practicable to measure an instantaneous value, so it has been decided
internationally that the tendency shall be taken as the change of pressure over the
preceding 3-hour period.

Some idea of the origin of these changes may be obtained by recalling the
meaning of pressure given at the beginning of this chapter. For a given place
it is seen that a change of atmospheric pressure can be produced only by a change
in the weight of the atmosphere and this, in turn, can be produced only by a net
addition or removal of air from a column of fixed cross-section above that place.
This may be brought about in two ways: firstly, by expansion or contraction
resulting from the heating or cooling of the atmosphere; and secondly, by means

of air passing into the column at a different rate from that at which it passes out. The first method is slow and can give only small tendencies of perhaps 1 millibar in 3 hours; the latter can be fast, giving large tendencies occasionally exceeding 10 millibars in 3 hours.

Lines may be drawn on a chart joining places which have equal values of tendency; such lines are *isallobars*. By analogy with isobars we have *isallobaric highs* and *lows*, showing centres of rising and falling pressure respectively (strictly, centres where pressure 'has risen' and 'has fallen' in the past 3 hours). They are useful in showing the direction of movement of isobaric features. For example, with an isallobaric low on one side of a depression and an isallobaric high on the opposite side, the depression moves (strictly, 'has moved') in a direction from isallobaric high to isallobaric low, that is from rising to falling pressure (see Figure 9). An anticyclone moves in the opposite direction.

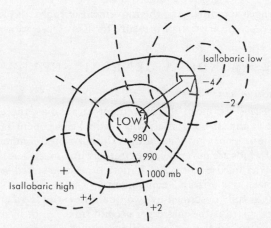

FIGURE 9. *Movement of a low indicated by positions of isallobaric centres*

Full and dashed lines represent isobars and isallobars respectively. Unit of isallobars: millibars per 3 hours.

When free from large-scale changes the barogram usually shows small, rhythmic, daily oscillations of 1 or 2 millibars with two maxima and two minima at very nearly the same times each day: maxima at about 1000 and 2200 local time and minima at about 0400 and 1600 local time. This *diurnal variation of pressure* is clearly related to the diurnal variation of temperature. This is to be expected, since at the time of maximum temperature the lowest layers of the atmosphere have their least density and so the pressure is least. Since a pressure fall implies a loss in mass above the observing point, the horizontal displacements which cause it must give corresponding accumulations of air elsewhere. As the earth rotates, the meridian of minimum pressure crosses any given place once each 24 hours. Now, by coincidence, it seems that the atmosphere has a period of oscillation of about 12 hours, so that the meridian of minimum pressure is accompanied by a second, though weaker, minimum on the opposite side of the earth and also passing around the earth once in 24 hours. The diurnal variation is thus tidal in origin—a *thermal* tide, not gravitational as are tides in the sea; it is most pronounced in equatorial regions but over Britain is usually masked by the

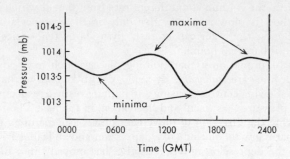

FIGURE 10. *Diurnal variation of atmospheric pressure at Kew*
Mean for years 1901–30.

much larger changes accompanying the movement of highs and lows. Figure 10 shows the mean diurnal variation of pressure for Kew Observatory.

2.2 WINDS—A GENERAL SURVEY

2.2.1 *Measurement of winds*

By the term *wind* we mean the movement of air over the earth's surface. This movement is almost horizontal; the vertical component is nearly always very small—in the order of one thousandth of the horizontal component. Although so small, vertical winds are nevertheless extremely important but their consideration will be delayed until Section 2.6. Unless otherwise stated we shall use the term 'wind' to mean its horizontal component.

Wind *velocity* is fully described by two quantities—its *speed* and its *direction*. The unit of wind speed is the metre per second (m s^{-1}) but in many countries, including Britain, two other units are still widely used, namely, the statute mile per hour (mile h^{-1}) and the nautical mile per hour or *knot* (kn). For approximate conversions it is useful to remember that:

1 metre per second is nearly 2 knots,
6 knots are very nearly 7 miles per hour.

An instrument for measuring wind speed is known as an *anemometer*; one which gives a continuous record (an autographic instrument) is an *anemograph*. There are many types but the principles of the commonest are:

(a) *Cup type*. Three or four cups are mounted symmetrically on a vertical axis which rotates at a speed largely dependent upon wind speed but independent of wind direction. Rotation occurs because the force exerted on the concave faces of the cups is more than on the convex faces.

(b) *Pressure-tube type*. When a tube, closed at one end, has its open end pointing into the wind the air entering the tube is brought to rest and creates inside the tube an excess pressure whose value increases with the wind speed. Also, when a tube, closed at both ends but with holes in its surface, lies with its axis perpendicular to the wind, as the air passes the holes it causes a reduction of pressure inside the tube the value of which increases with the wind speed.

(c) *Pressure-plate type.* A plate is allowed to hang from a horizontal axis normal to the wind. The inclination of the plate to the vertical increases with the wind speed.

(d) *Hot-wire type.* A piece of platinum wire is heated electrically to a temperature much above that of the passing air. The current intensity required to maintain a constant wire temperature increases with the wind speed because the stronger the wind the greater the rate of loss of heat by conduction.

Wind direction is the direction *from* which the wind blows, measured from true north either clockwise in degrees or as a compass point. An instrument for measuring wind direction is a *wind vane*, a body mounted unsymmetrically and free to turn about a vertical axis; it takes up a position so that the direction of the force on it, as a result of the wind's horizontal pressure, passes through the vertical axis, so making the centre of pressure lie to the leeward of the axis.

It is important to distinguish between two terms used when describing changes of wind direction; they are *backing*, meaning a wind direction changing anti-clockwise, and *veering*, a clockwise change.

2.2.2 *Some properties of surface wind flow*

It is well known that the wind velocity, both speed and direction, at any given place is continuously changing. We would expect, therefore, that the velocity of any parcel of air as it moves over the earth's surface would also be continuously changing. This, indeed, is found to be true if we follow the path (*trajectory*) of such a parcel using wind measurements from a large number of neighbouring observers. Now a change in velocity, that is an *acceleration*, requires a force to bring it about according to Newton's Second Law of Motion. We are led to believe, then, that a force (or forces) acts on the parcel of air causing it to move. When these forces just balance each other there is no acceleration and the parcel moves with a constant velocity—we then have *balanced flow*. When the forces do not balance, the parcel accelerates.

Experience shows there are two main types of surface wind velocity change:

(a) Irregular, short gusts and lulls, lasting only a few seconds at most.
(b) Smooth, long-term changes, usually occurring gradually over a period of hours or even days.

These two types can usually be seen on an anemogram as, for example, the one shown in Figure 11. The former are interesting but ignored when we speak of 'the wind', meaning a mean wind over a 10-minute period, say. The latter changes take place so slowly that the accelerations are small, so small that we may consider that normal wind flow is nearly balanced. Later we will consider the nature of the forces causing wind flow and how they control its velocity.

2.2.3 *Buys Ballot's Law*

An interesting relationship exists between the observed wind velocity and the pattern of isobars in the vicinity of the point of observation. Figure 12 shows a chart covering part of the northern hemisphere with isobars drawn and with winds plotted by using 'wind arrows' in the conventional way. Each wind observation is represented by a 'shaft', drawn from the observing point in the direction from which the wind is coming, and 'feathers' to indicate speed, one whole feather

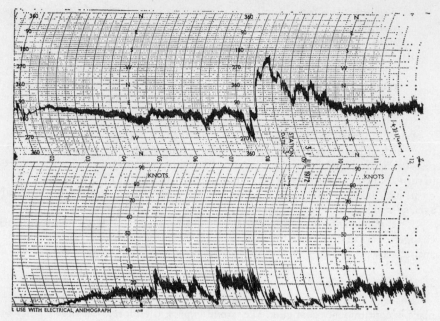

FIGURE 11. *Section of an anemogram showing both short- and long-term variations of surface wind velocity*

representing 10 knots and half a feather 5 knots. It is apparent immediately that:

(a) the wind blows nearly parallel to the isobars with lowest pressure to the left of the direction of motion,

(b) the wind speed is largest where the isobars are most closely spaced.

A relationship between the direction of wind and isobars was first put into the form of a law by Buys Ballot in 1857. *Buys Ballot's Law* states that if you stand with your back to the wind, the lowest pressure lies on your *left* in the *northern*

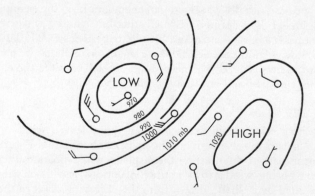

FIGURE 12. *Relation between isobars and winds*

hemisphere but on your right in the southern hemisphere. It is only approximately true; nevertheless, it is a useful guide. It follows from the law that, in the northern hemisphere, the winds form a clockwise circulation around an anticyclone but anti-clockwise around a depression.

2.3 GEOSTROPHIC WIND

2.3.1 *Controlling forces*

We have seen that wind flow is usually nearly balanced. To understand which forces bring about air motion and how they control it, let us simplify the problem by assuming, firstly, that wind flow is *exactly balanced* and secondly, that only *two* forces are acting on any given parcel of air, causing it to move horizontally at a *constant speed*. Vertical forces, such as the parcel's weight, need not be considered because they do not affect the horizontal motion. In Section 2.4 we will consider some of the consequences of these assumptions.

We already have some knowledge of one of the two forces which we are assuming produce the motion: it results from horizontal variations of the atmospheric pressure near the parcel. If the pressure is low on one side and high on the opposite, a force tends to make the parcel move in a direction from high to low pressure. With close isobars pressure varies rapidly along a line normal to them. A measure of the intensity of this variation is the *pressure gradient*, G, or the rate of change of pressure with distance normal to the isobars; it is usually in the order of $0 \cdot 01$ mb km^{-1}. A large (steep) pressure gradient results from close isobars; a small (slack) gradient results from widely spaced isobars. The force caused by these pressure variations is known as the *pressure gradient force*, X; its size is directly proportional to the size of the pressure gradient, G, that is,

$$X = k_1 \times G, \qquad \qquad \ldots \ldots (2)$$

or inversely proportional to the spacing of the isobars. In direction, it always acts normal to the isobars from high to low pressure. It is clearly zero when there is no gradient, that is when pressure is uniform.

If the pressure gradient force were the only force acting, the air flow would be unbalanced and the air would tend to rush inwards towards the low-pressure centre giving an accumulation of air there and a consequent rise of pressure until the low disappeared. This is not observed in practice so there must be at least one other force present which can balance the pressure gradient force. The most important of these is a consequence of the rotation of the earth. This rotation means that an observer standing still on the surface of the earth in the northern hemisphere is himself rotating anti-clockwise about the local vertical with a spin which is greatest at the north pole where it is equal to the spin of the earth about its axis and which decreases to zero at the equator. To such an observer a particle which is in fact moving in a straight line appears to be continuously turning to the right of its direction of motion. This deflection implies that the particle has an acceleration and calculation shows that it acts at right angles to the direction of motion. The force required to cause this acceleration is known as the Coriolos force, Y; its magnitude is proportional to the speed, V, of the particle, that is

$$Y = k_2 \times V, \qquad \qquad \ldots \ldots (3)$$

and it always acts perpendicularly to the direction of motion of the particle, to the right in the northern hemisphere, to the left in the southern hemisphere. It is clearly zero when V is zero.

2.3.2 *Properties of the geostrophic wind*

A wind which blows under the influence of these two forces alone, and which is balanced, is known as the *geostrophic wind*. We can deduce its properties from our knowledge of the two forces. Figure 13(*a*) represents a small portion of a chart in the northern hemisphere on which two isobars have been drawn near an air parcel, A. Let us consider:

(a) The *direction* of the resulting wind. Because of the orientation of the isobars, the pressure gradient force must be as shown and, since there is balanced flow, the Coriolis force must be equal in size and opposite in direction to it, that is, towards high pressure. From this it follows that the parcel must be moving in the direction shown if the Coriolis force is to act to the right of its direction of motion. Hence, the parcel moves exactly parallel to the isobars with lowest pressure to its left.

(b) Its *speed*. Since the pressure gradient force just equals the Coriolis force, from equations (2) and (3) it follows that

$$k_1 \times G = k_2 \times V,$$

that is,
$$V = k_3 \times G,$$

showing that the speed is directly proportional to the pressure gradient. This means that the speed is inversely proportional to the isobar spacing: close isobars, strong wind; widely spaced isobars, light wind.

It should be carefully noted that the geostrophic wind is an idealized, theoretical wind; it cannot be measured but it can be calculated for any given place so long as the isobar pattern around that place is known. Given both scale of chart and isobar interval (usually 2 or 4 millibars) each value of the distance taken normal to two adjacent isobars corresponds to a certain geostrophic wind speed. Halving the distance doubles the wind speed. A *geostrophic scale* can be devised to evaluate the geostrophic wind speed by measuring the isobar spacing.

An interesting point arises from the value for k_2 (known as the *Coriolis parameter*). It can be shown that

$$k_2 = 2m\omega \sin \varphi$$

where m is the parcel's mass, ω is the angular velocity of rotation of the earth (once per day or 24 h^{-1}) and φ is the latitude. On the equator, where $\varphi = 0$, $k_2 = 0$, that is, there can be no Coriolis force on the equator. It follows from this too that, for a given isobar spacing, the geostrophic wind speed is least near the poles.

2.4 OBSERVED WINDS NEAR THE GROUND

2.4.1 *Effects of friction*

The surface wind velocity is normally measured at a standard height of 10 metres (33 feet) above the ground. Outside the tropics if we compare a measured

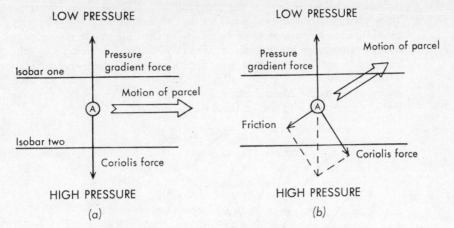

FIGURE 13. *Balanced wind flow involving (a) pressure gradient force and Coriolis force only, (b) pressure gradient force, Coriolis force and friction*

surface wind with the geostrophic wind calculated from a sea-level pattern of isobars, we find a similarity but, at the same time, we also find at least two important differences:

(a) The observed *direction* is backed from the geostrophic direction by an amount which is rather variable, approximately 30° over land and 10° over the sea.

(b) The observed *speed* is less, about one-third of the geostrophic speed over land and two-thirds over the sea.

These differences are caused by another force which we have ignored so far, namely, *friction*—both between the air and the ground and internally within the air itself. When the three forces, pressure gradient force, Coriolis force and friction, just balance, the wind blows as in Figure 13(*b*). Now the effect of friction is most marked at the earth's surface and it decreases rapidly upwards, becoming negligible above a certain height which is rather variable but on average about 500 metres. This level marks the maximum vertical extent of the irregular motion, known as turbulence, induced by wind flow over the rough ground. It is to be expected then, that the greatest departure of the observed wind from the geostrophic value should occur just near the ground and that, as we ascend, this departure should decrease until above about 500 metres the observed wind should be almost geostrophic. Measurements show this to be true, in general, and it explains the necessity for exposing anemometers and vanes at a standard height if readings from neighbouring stations are to be compared.

2.4.2 *Other effects*

Friction is very important at all times in modifying wind flow but there are still further effects to be considered, mostly of a temporary or a localized nature. These can alter the observed wind so as to make it bear little or no relation to the geostrophic wind. We will consider a few of these effects.

(a) Wind blowing around obstacles, such as houses, trees or hills, forms turbulent eddies whose sizes are comparable with those of the obstacles. They greatly alter the local wind so that an anemometer must be very carefully sited in an open place to minimize their effect. On a very large scale, a valley often has light winds when the isobars lie across its length; but sometimes the winds may be excessively strengthened—*funnelling*— if the isobars run along the valley. Funnelling helps to produce such valley winds as the Mistral along the Rhône (see Figure 14). Even our British mountains give similar local winds.

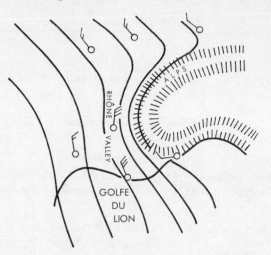

FIGURE 14. *Funnelling down the Rhône Valley producing the Mistral*

(b) During our consideration of the geostrophic wind in Section 2.3, we assumed several conditions without stating their nature. On some occasions these assumptions are not valid with the result that the wind for balanced flow differs from the geostrophic wind. The most important of these effects arises from curvature of the path, or trajectory, of an air parcel flowing around an anticyclone or a depression. A curved path can be followed only if there is an acceleration towards the centre of curvature, that is, only if there is a net force acting inwards to the centre of curvature. For anticyclonic curvature (Figure 15(a)) the Coriolis force must be greater than the pressure gradient force with the result that the wind speed for balanced flow is *super*-geostrophic; but for cyclonic curvature (Figure 15(b)) the Coriolis force is less than the pressure gradient force and the balanced flow is *sub*-geostrophic. Balanced flow along a curved path under the influence of these two forces only is known as the *gradient wind*; the greater the curvature, the greater the difference between corresponding gradient and geostrophic winds. Apart from curvature of the isobars, we may just note some other effects arising either from the isobars not being parallel (wind blowing towards closer isobars is sub-geostrophic), or from their not being parallel to the equator (k_2 increases polewards, therefore poleward wind flow is super-geostrophic), or from their movement (an 'isallobaric component' is added, blowing towards the isallobaric low, normal to the isallobars— see Section 2.1.4).

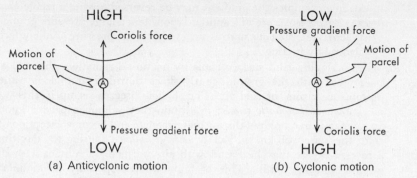

FIGURE 15. *Balanced wind flow along curved isobars (gradient wind) in the northern hemisphere*

(c) When the geostrophic wind is weak (say less than about 10 knots), a local *sea-breeze* may be established near a coastline. This takes place in the following way. With clear skies the day-time land temperature rises very much more than that of the sea and so the air temperature over the land much exceeds that over the sea. Vertical expansion causes pressure gradients to form at higher levels, above the surface, because more air will lie above a given level after the expansion than before, and the resulting rise in pressure at that level over the land will cause a gradient such that the air flows gently from the land towards the sea. The removal of this air causes a pressure fall over the land at sea level and an accompanying rise over the sea, that is, a pressure gradient develops normal to the coast and sea air starts to flow inland. At first, the Coriolis force hardly influences the breeze's velocity, a balance being obtained between the pressure gradient force and friction; but after some hours its effect increases progressively so that the breeze tends to veer towards a truly geostrophic direction with low pressure on the left (landwards). Air from over the sea spreads inland for many miles as a breeze of up to 10 to 15 knots displacing land air which returns seawards above it (Figure 16). Its furthest penetration inland is often marked by a line of horizontal convergence and vigorous convection (see Section 2.6.3), sometimes termed a 'sea-breeze' front. This line of rising air is much favoured by glider pilots seeking to gain height. The sea-breeze dies away at night when the pressure gradient is destroyed by rising pressure over the land caused by radiational cooling;

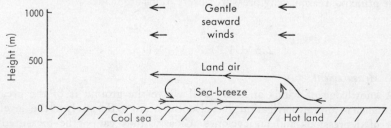

FIGURE 16. *Diagrammatic cross-section through a well-developed sea-breeze*

in extreme examples the gradient may be reversed, giving a feeble *land-breeze*. *Ravine winds* are of a similar type where a large pressure gradient exists across the ends of a narrow valley. Again the pressure gradient force is balanced by friction.

(d) When a sloping land surface cools by nocturnal radiation, the cold air in contact with the ground gently flows downhill and along valley bottoms as 'rivers' of cold air. The shallow breezes set up in this way are known as *katabatic winds*; in Britain they seldom exceed 5 knots. In some favoured hollows the air may stagnate and become exceptionally cold; these are well known locally as *frost hollows*. During the day-time a reverse breeze may be found with air, warmed by contact with the ground, flowing uphill. This is found more especially in mountain regions where one side of a valley may be heated much more than the other. These are *anabatic* or *mountain winds*. Along a mountainous coastline, sea- and land-breezes are combined with katabatic and anabatic winds.

2.4.3 *Some details of surface wind flow*

The *diurnal variation* of the surface wind is largely controlled by diurnal changes of stability of the turbulent layer near the ground. In stable air, as may be found during a clear night, mixing is inhibited and the frictional force is large; but when it is unstable, mixing is extensive and friction is reduced. Maximum speed therefore occurs in the early afternoon, and the minimum in the early morning. Correspondingly, the direction is least backed in the afternoon and most in the early morning.

The wind nearly always shows some *gusts*. These are caused by:

(a) Turbulence; when they are on a small scale, with extreme gust speeds approximately in excess by one-third above the mean speed.

(b) Convection; when they are very irregular in size and timing and there is a marked diurnal variation, as is to be expected, with the strongest gusts in the afternoon.

Squalls last for several minutes at least and are associated with large convective clouds or fronts—see Sections 7.3.1 and 10.4.2. A *gale* is just a wind whose mean speed exceeds 34 knots (see also Section 10.1.3). For *hurricanes*, see Section 10.4.

As long as all these possible modifications are borne in mind, a chart on which isobars have been drawn may be used to estimate surface wind velocities in those areas where direct observations are not available. This is, in fact, one of the principal reasons why pressure observations are made.

2.5 UPPER WINDS

2.5.1 *Measurement*

A knowledge of winds at levels well above the ground is of the greatest importance to aircraft navigation; it is also important to the study of the development of depressions and anticyclones. Clearly, they cannot be measured by anemometers and wind vanes, except perhaps in the first 300 metres above the ground where the instruments may then be mounted on a tower. Other methods must be used; of these, we may mention:

(a) Noting the variation in position of a hydrogen-filled balloon which rises freely. Its position may be fixed accurately if three quantities are known: azimuth (bearing from true north), elevation (above horizon) and height, all measured relative to the point of release. The first two can be found either optically, by using a theodolite, or by radar, which is obviously preferable in cloudy weather. The height is found either by knowing the rate of the balloon's ascent (obtained by putting a known mass of hydrogen in the balloon) or by a direct measurement of the distance of the balloon using radar. By measuring these three quantities, say every minute, mean winds are obtained for each successive minute-long interval, that is, for progressively higher layers in the atmosphere. For a balloon ascending at the rate of 300 metres per minute, mean winds in 300-metre layers may be found until either the balloon bursts or it is lost. In this way, winds up to 20 or 30 kilometres (say, 60 000 to 100 000 feet) are now being measured regularly over large areas of the world. Quite often, radiosonde balloons are used for this purpose.

(b) Noting the drift of an aircraft from its set course.

(c) Observing the motion of a piece of cloud (which must, of course, be moving with the wind—not all clouds do this).

(d) Using meteorological satellites (see Appendix B, Section B.2.4).

2.5.2 *Upper level charts*

Just as we plot values of pressure at mean sea level (MSL) on a surface chart and then draw isobars, so we could plot values of pressure at any other level above MSL and draw isobars for that level. The data could be obtained from radiosonde ascents and such a chart would be a constant-level chart. This type of chart was drawn at one time but nowadays we work with constant pressure charts. Instead, then, of plotting pressure at a selected height above MSL and drawing isopleths of pressure (isobars), we plot values of height above MSL of the selected pressure surface and draw isopleths of this height (contours). The pressure levels commonly in use are 700, 500, 300, 200 and 100 millibars, corresponding in middle latitudes to heights of about 3, $5\frac{1}{2}$, 9, 12 and 16 kilometres respectively above MSL.

The height at which a given pressure is found above a particular place may be calculated from temperature data obtained by a radiosonde ascent made at that place. The atmosphere between the pressure level and the ground is divided into a convenient number of layers, each, say, 100 millibars thick, and the depth of each layer is found by using equation (1), where p is now 100 millibars, h is the depth and ρ is the mean density of the layer, found from the vertical temperature distribution. By adding the depths of each layer the height of the pressure level is found. The reason why several layers are taken is that it is more difficult to calculate a mean density for one deep layer than for each of several thin layers.

Figure 17 is an example of a contour chart for the 300-millibar level with contours drawn at 60-metre intervals. Winds measured at 300 millibars have also been plotted and a now-familiar relationship reappears: the winds blow parallel to the contour lines with lowest height on the left in the northern hemisphere, and their speed is directly proportional to the contour gradient.

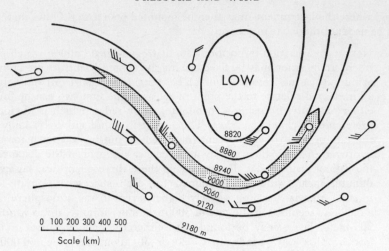

FIGURE 17. *Example of a contour chart for* 300 *millibars, showing contours at*
60-*metre intervals*

The shaded area represents the axis of a jet stream.

Once again we can define a geostrophic wind but this time using the *contour*
pattern to calculate it. The terms 'high', 'low', etc., may still be used when
describing contour patterns but, of course, with reference to height above mean
sea level, not pressure.

A constant-pressure chart thus represents a surface which is not horizontal
but clearly the contours, being lines of constant height are themselves horizontal
and since the geostrophic wind is parallel to the contours, it too is horizontal as it
was defined as being on a constant level chart. Note that a low (or a high) in a
contour pattern will correspond very closely with a low (or a high) in the isobaric
pattern on a constant level chart at approximately the same level. For example,
in Figure 17 the height of the 300-millibar surface above MSL varies between 9180
metres and 8820 metres, i.e. the average height is about 9 kilometres. Had we
drawn the isobars on the 9-kilometre constant level chart, the 300-millibar isobar
would have coincided exactly with the 9000-metre contour. Where we have the low
the 9-kilometre surface would be at its maximum height (9000 − 8820 = 180 m)
above the 300-millibar surface and hence the pressure at 9 kilometres would attain
its greatest depression below 300 millibars, i.e. we should have an isobaric low
corresponding to the contour low.

2.5.3 *The jet stream*

Aloft, wind speeds are usually much greater than those observed near the
ground. In fact, the wind over Britain usually increases with height up to near
the tropopause, and especially is this true of its component from the west. The
strongest winds may exceed 100 knots, and very occasionally 200 knots, especially
in winter. However, at any given level these strong winds are often confined to
a definite narrow band perhaps 150 to 500 kilometres wide, an example being shown
in Figure 17. They are even more restricted in the vertical being only a few

kilometres deep. These belts of strong wind are known as *jet streams*, a term introduced by Rossby in the United States of America in 1947. They may be a thousand
miles or more in length with the winds blowing along their length. A jet stream
is usually slow moving so that the air blows through it, that is, a jet has one end
as an entrance and the other as an exit. The strongest winds are found just below
the tropopause; both above and below the jet stream and on either side of it
wind speeds rapidly decrease, so that the jet stream may be considered as a
ribbon of strong winds embedded in lighter winds.

   More will be said about the jet stream when we come to consider the three-
dimensional structure of a depression in Chapter 10.

2.5.4 *The thickness chart*

   In general, at any given time the patterns of contours for different pressure
levels resemble each other, but there are always some differences. In an extreme
example, a low at 700 millibars may be surmounted by a high at 300 millibars,
with a corresponding difference in the wind patterns at these levels. It follows,
then, that the wind velocity always varies with height above a given place,
sometimes markedly, and it is useful to understand why this should be.

   First, we will consider another type of upper-air chart—the *thickness chart*.
The difference between the heights of two pressure levels at any given place is
known as the *thickness* of the layer of the atmosphere between those two levels.
Its size increases with the *mean* temperature of the layer: large thickness, warm
layer; small thickness, cold layer. This can be seen by considering a vertical air
column lying between the two pressure levels and then using a modified form
of equation (1):

$$p = \rho g t, \qquad \ldots\ldots (4)$$

where p is now the pressure difference between the bottom and top of the
column, ρ is the mean density and t is the height of the column, that is, its thickness.
Rewriting equation (4):

$$t = \frac{p}{\rho g} \qquad \ldots\ldots (5)$$

and, since ρ is inversely proportional to the mean absolute temperature T, that is,

$$\rho = \frac{k}{T}$$

where k is a constant, then equation (5) becomes

$$t = \frac{pT}{kg}. \qquad \ldots\ldots (6)$$

Since p, k and g are constant for a given place and layer, equation (6) shows
that the thickness of a layer is proportional to its mean temperature.

   A chart on which values of the thickness of a given layer are plotted is known
as a thickness chart. Lines may be drawn upon it joining places of equal thickness
and they are known as *thickness lines* (sometimes, *relative contours*). It is important always to remember to which layer in the atmosphere the lines refer; the
usual thickness chart is for the layer 1000 to 500 millibars. Some of the lines on
such a chart will be complete curves enclosing areas of relatively low or high

values of thickness, thus indicating areas of relatively low or high mean temperature, respectively, of the layer concerned. A centre of low thickness is known as a *cold pool*; similarly we have *warm pools*, *cold troughs* and *warm ridges*. A thickness chart shows at a glance where the coldest or warmest air lies in the layer, and this is one if its major uses. The boundaries of neighbouring masses of air with markedly different temperatures are also clearly shown, since it is at these boundaries, the frontal zones, that the thickness lines are closest together.

Now we can see that the contour patterns at two pressure levels will be different if the mean temperature between the levels varies from place to place. Consider an area of the 1000-millibar level where there are no contour lines, that is, where the height of the pressure level is constant. Suppose above one part of this area the atmosphere is cold up to 500 millibars and above another part it is warm. In the cold part, the 500-millibar level will be comparatively low and in the warm part it will be high, so that a contour gradient will exist at 500 millibars. With no contours at 1000 millibars there will be no 1000-millibar wind, but with the contours existing at 500 millibars there will be a 500-millibar wind, and this difference depends solely upon the horizontal variations of mean temperature in the layer between 1000 and 500 millibars.

2.5.5 *The thermal wind*

It has been stated that the wind velocity changes with height; we describe this as vertical *wind shear*. It follows from Section 2.5.4 that the cause of this vertical wind shear above any given place is the uneven distribution of mean temperature in the horizontal. The *vector* difference between the winds at the top and bottom of any layer is known as the *thermal wind* of that layer, making sure that the lower wind is subtracted from the upper. It may be found graphically by using a simple construction as in Figure 18(*a*) where V_l and V_u represent the lower and upper wind respectively, and the thermal wind is as shown; note its direction, from the tip of the V_l arrow to the tip of the V_u arrow. Like the geostrophic wind, the thermal wind is also theoretical; it cannot be measured but may be calculated from a knowledge of the winds at the top and bottom of the layer.

If, on a thickness chart for a given layer, we plot thermal winds for the same layer, we again find a relationship between the winds and the lines drawn. The thermal wind lies parallel to the thickness lines with lowest thickness (that is, lowest mean temperature of the layer) to its left in the northern hemisphere,

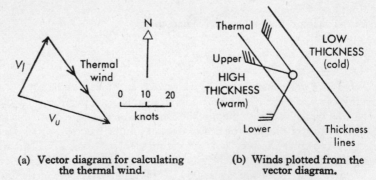

(a) Vector diagram for calculating   (b) Winds plotted from the
     the thermal wind.   vector diagram.

FIGURE 18. *The thermal wind*

and its speed is directly proportional to the thermal gradient, that is, inversely proportional to the spacing of the thickness lines. This relation gives us one of the chief uses of thermal winds—helping to draw thickness lines where direct measurements of the thickness itself are not available. The winds of Figure 18(a) are plotted in Figure 18(b), and from the above relation we can deduce the orientation of the thickness lines around the observer—cold air lies to the left, that is, to the north-east, and warm air to the right, that is, to the south-west.

In Figure 18(b) we see that winds at both the bottom and top of the layer have components from the direction of the warmer air, so the warmer air must be moving in such a way that it is replacing the colder air. It is left as an exercise to show that this is always so if the wind veers with height. As a rule of thumb, a wind *veering* with height indicates that *warmer* air is approaching (warm advection), and conversely a wind *backing* with height indicates *cold* advection. This rule illustrates another important use of thermal winds and may be employed by a surface observer who notes the wind velocities at two different levels by watching the motion of clouds. In Section 10.1.4 we will see how to use the rule when discussing the patterns of winds above a depression.

2.6 VERTICAL WINDS

2.6.1 *Flow over an obstacle*

So far, only *horizontal* winds have been discussed. All natural winds have some *vertical* component, however, and we will now consider their origin, magnitude and significance. There are three main causes of vertical motion: flow over an obstacle, convection and convergence. In Section 2.4.1 we studied the effects of friction on the wind as it flows over rough ground and we saw that it causes turbulence, which churns up the lowest 500 metres or so of the atmosphere. The effect of the turbulent eddies is to superimpose small updraughts and downdraughts on the general wind, and although these vertical currents may sometimes be comparable in magnitude with the horizontal wind, they are quite localized. An aircraft flying through the turbulent air experiences a succession of jolts, which rarely prove dangerous.

When the wind blows over a mountain, the atmosphere is displaced upwards bodily and to an extent which may still be significant at levels many times the height of the mountain. This *orographic ascent* may be widespread when air flows across large mountain ranges. A parcel of air at the crest of its displacement will be subjected to buoyancy forces tending to restore it to its original level, as long as the atmosphere is stable. Only in the lee of the mountain is it able to return, but the downdraught usually overshoots its original level, finally setting up oscillations in the form of a set of standing waves, or *lee waves*, downwind from the mountain (see Figure 19). The greater the stability of the atmosphere, or the weaker the wind flow, the shorter is the wavelength; the amplitude increases with the height of the mountain. Lee waves are almost stationary relative to the mountain, and with strong winds may contain powerful updraughts and downdraughts which can exceed 20 knots. Glider pilots have found them very useful for gaining lift but they can be dangerous to powered aircraft if they are encountered unexpectedly. Below some of the crests the surface wind may be reversed, giving a *rotor* (Figure 19), and on the leeward side of the mountain the downward-flowing air may give a local strong wind—the *helm wind*.

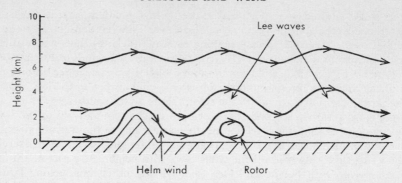

FIGURE 19. *Airstream flowing over a hill producing lee waves, a rotor and helm wind*

2.6.2 *Convection*

If the atmosphere is unstable, bubbles of warm, buoyant air rise spontaneously with vertical speeds in the order of 2 to 10 knots. In extreme conditions the updraughts may exceed 30 knots and can be very dangerous to aircraft. They are often associated with particular cloud forms which can thus be used to indicate their presence, and similar localized but powerful downdraughts may occur next to them.

For further details see Section 5.4.

2.6.3 *Convergence*

When there is a net inflow of air into a region, we say that the air motion is convergent or that *convergence* is occurring; with a net outflow from a region the air motion is divergent—*divergence* is occurring. Any marked convergence into a region would result in an accumulation of air leading to a much greater increase of density than is ever observed to occur. In fact air motion in the atmosphere is approximately non-convergent (non-divergent). However, if we consider only the horizontal motion of the air we do find regions where marked convergence or divergence is occurring. Very little change in density results because convergence (divergence) of the horizontal motion is almost exactly balanced by divergence (convergence) of the vertical motion. Thus convergence of horizontal motion near the ground is accompanied by upward motion whereas convergence of horizontal motion in mid troposphere may be accompanied by upward or downward motion or both and convergence of horizontal motion near the top of the troposphere usually by downward motion, since the stability of the stratosphere is not favourable to any vertical motion. With divergence of the horizontal motion, the sense of these vertical motions is reversed. In future, when we use the words convergence or divergence, we shall always understand them to refer to the horizontal component of the wind only.

Developing depressions and troughs of low pressure are favourable regions for convergence in the lower troposphere and divergence aloft, whilst developing anticyclones and ridges of high pressure are favourable regions for divergence below and convergence aloft. These ideas lead us to the broad generalization that developing lows are associated with ascending air and highs with descending air. The vertical motion produced by convergence and divergence is almost always very weak, only a few feet per minute, and unimportant from the point of view of the aviator; however, its weakness is offset by its persistence. Air flowing through an area of convergence, which is often very extensive, covering thousands of square kilometres, may take many hours and during this time it can ascend or descend several thousand feet. In Section 5.2.1 we shall see that upward and downward motion, with its accompanying adiabatic expansion and compression, is responsible for the formation and dispersal of most of our clouds. Hence, we can understand the commonly held ideas that lows are predominantly cloudy and highs cloudless. However, there are many exceptions, some of which will be dealt with in Part II of this book.

2.6.4 *Frontal ascent*

In Section 1.6 we saw that a frontal zone is a boundary separating air masses of different origin (and therefore of different properties, particularly temperature). Such boundaries, especially when associated with depressions, are often regions of marked convergence in the lower troposphere with, in general, the warm air rising above the cold. This frontal ascent is often widespread and persistent, and is usually accompanied by correspondingly large areas of cloud and precipitation.

For further details see Section 9.6.

BIBLIOGRAPHY

AANENSEN, C.J.M.; 1965. Gales in Yorkshire in February 1962. *Geophys Mem*, 14, No. 108.

COULMAN, C.E.; 1970. Fluctuations of horizontal wind speed near the ground in convective conditions. *Q J R Met Soc*, 96, pp. 529–534.

CORBY, G.A.; 1957. Air flow over mountains. *Met Rep*, 3, No. 18.

GUNN, D.M. and FURMAGE, D.F.; 1976. The effect of topography on surface wind. *Met Mag*, 105, pp. 8–23.

LAWRENCE, E.N.; 1954. Nocturnal winds. *Prof Notes Met Off*, 7, No. 111.

McGINNIGLE, J.B.; 1963. Katabatic winds at Acklington during a very cold spell. *Met Mag*, 92, pp. 367–371.

PEDGLEY, D.E.; 1974. Field studies of mountain weather in Snowdonia. *Weather*, 29, pp. 284–297.

SAMPSON, J.E.; 1964. Sea-breeze fronts in Hampshire. *Weather*, 19, pp. 208–220.

SHELLARD, H.C.; 1967. Wind records and their application to structural design. *Met Mag*, 96, pp. 235–243.

SHELLARD, H.C.; 1975. Lerwick anemograph records 1957–70 and the offshore industry. *Met Mag*, 104, pp. 189–208.

STUBBS, M.W.; 1975. An unusually large fall of pressure. *Weather*, 30, pp. 91–92.

WALLINGTON, C.E.; 1960. An introduction to lee waves in the atmosphere. *Weather*, 15, pp. 269–276.

WARD, F.W.; 1953. Helm-wind effect at Ronaldsway, Isle of Man. *Met Mag*, 82, pp. 234–237.

WATTS, A.; 1967. The real wind and the yacht. *Weather*, 22, pp. 23–29.

CHAPTER 3

WATER IN THE ATMOSPHERE

3.1 SOME PHYSICAL PROPERTIES OF WATER

3.1.1 *Phase changes*

Like most substances, water exists in three *phases*, or *states of matter*—solid, liquid and vapour. The first two, *ice* and *water*, are well known but the significance of the third, *water vapour*, can be easily overlooked because of its colourless and transparent nature. However, water vapour is an important constituent of the atmosphere, and although it is present in small quantities only—less than 4 per cent by weight—without it we would have no clouds or rain, temperature ranges would be more extreme and, in general, meteorology would be far less interesting. It is useful, therefore, to know a little of those properties of water which are of importance in understanding the physical processes taking place in the atmosphere.

The three phases of a substance can be changed reversibly into each other, and since all of the six possible transformations are of interest to us, we must distinguish carefully between them.

Initial phase	Final phase	Name of phase change
Solid	Liquid	Melting, or fusion
Liquid	Solid	Freezing, or crystallization
Liquid	Vapour	Evaporation
Vapour	Liquid	Condensation
Solid	Vapour	Sublimation
Vapour	Solid	Deposition

Consider these changes applied to water. Under ordinary conditions ice melts at a certain fixed temperature, 0 °C, known as its *melting-point*. Below 0 °C ice is the stable phase, and above 0 °C water is the stable phase. Although it seems that for practical purposes ice cannot exist at temperatures above 0 °C, nevertheless water can exist in the liquid phase at temperatures below 0 °C, even as low as −40 °C. In such a state it is said to be *supercooled*. Supercooled water is a metastable phase, that is, although it is stable it is not the most stable phase for temperatures below 0 °C, and it has a great tendency to change into ice. Thus, when water is cooled, it need not freeze at 0 °C, and so the *freezing-point* of water is not fixed but its maximum value is 0 °C of course. The frequent occurrence of supercooled water in the atmosphere makes it very important and it will be considered further in Section 3.4.1. Supercooled water cannot exist in contact with ice, so that when the two are brought together the water freezes. However, there is evidence to support the idea that an ice surface has a structure resembling that of water even at temperatures well below 0 °C—see also Section 6.2.5.

Phase changes involving water vapour (evaporation, condensation, sublimation and deposition) can occur at temperatures up to a certain maximum, known as the *boiling-point*. Below this temperature water vapour can exist in contact with liquid water (or ice, if below 0 °C), but above the boiling-point only the

42

vapour is the stable phase; water whose temperature is higher than the boiling-point is said to be *superheated*, but in meteorology this condition is of little interest.

The numerical values of both the melting-point of ice and the boiling-point of water depend upon the pressure exerted upon their surfaces. For ordinary conditions, when this pressure is exerted by the atmosphere, ice melts at 0 °C and water boils at 100 °C. An increase of pressure decreases the melting-point but the effect is small because an increase to 132 atmospheres is needed to reduce the melting-point to −1 °C. This property of ice is employed when one makes snowballs by compressing freshly fallen snow, and it is also illustrated by the formation of a crust of snow on the soles of one's shoes when walking on fresh snow. In both examples, compression results in localized melting of the ends of pointed ice crystals in the snow where intense pressures are set up on contact. When the pressure is removed, the resultant water freezes and rigidly joins the neighbouring crystals. If the snow is originally very cold, the binding produced by compression is small because the melting-point cannot be lowered below the existing temperature.

The change of boiling-point with pressure will be discussed in Section 3.1.4.

3.1.2 *Latent heat*

Heat is absorbed when a substance melts or evaporates, even though the temperature remains constant. This heat which is absorbed at constant temperature is known as *latent heat*. The amount of heat needed to melt 1 kilogram of a substance at constant temperature is known as the latent heat of fusion—for ice at 0 °C it is 334×10^3 J. Similarly the latent heat of vaporization is the amount of heat needed to evaporate 1 kilogram of a substance also at constant temperature—for water it is 2406×10^3 J at 40 °C but 2501×10^3 J at 0 °C. The latent heat of sublimation of ice is 2834×10^3 J kg^{-1} at 0 °C. When a phase change is reversed the latent heat is released again.

If there is no external source of heat when evaporation takes place, the latent heat of vaporization must be supplied from the kinetic energy of the molecules within the liquid. A physical picture of this process may be seen by considering evaporation as the removal of the fastest-moving molecules from the liquid's surface—a removal of those molecules which can best escape from the attraction of their neighbours by virtue of their greater kinetic energy. After their escape, these fast molecules leave behind a liquid containing molecules whose mean speed, and therefore kinetic energy, is less than it was before evaporation occurred. Since the temperature of a substance increases with the kinetic energy of its constituent molecules, evaporation, because it is accompanied by a decrease in the kinetic energy of the liquid's molecules, must cause a fall in the temperature of the liquid. Therefore, evaporation results in cooling; conversely, condensation results in heating—similarly for sublimation and deposition.

The need for latent heat arises from the overall rearrangement of the molecules of the substance when a phase change occurs. In the solid phase the positions of the molecules are fairly fixed relative to each other, but in the liquid they are more mobile and their movements are partly independent. This mobility is brought about by breaking some of the inter-molecular bonds in the solid, and since this breakage requires work, some energy is used up in the form of heat. In the vapour the molecular spacing is relatively large, so that the increased separation accompanying evaporation or sublimation requires even more work to break the bonds originally present. Hence, the latent heat of vaporization is greater than the latent heat of fusion. The value of the latent heat of vaporization decreases as the temperature increases, because at higher temperatures the inter-molecular spacing in the liquid is greater and the phase change can occur more easily.

3.1.3 *Density*

The density of a solid or liquid usually decreases as the temperature increases because the inter-molecular spacing gradually increases with the temperature. Ice is normal in this respect: its density is 923 kg m^{-3} at $-40\ °C$ and 917 kg m^{-3} at $0\ °C$.

At the melting-point the density of the liquid phase is usually less than that of the solid because fusion is accompanied by the breakage of bonds in the solid, thus causing a sudden increase in the spacing of the molecules. Water is anomalous in this respect since the density of water at $0\ °C$ is greater than that of ice, being 999·84 kg m^{-3}. This is because the structure of ice contains 'holes', some of which become filled by mobile molecules on fusion. Water is also anomalous in that it has a maximum density at $4\ °C$ (strictly, $3·98\ °C$) of 999·97 kg m^{-3}. This peculiar change of density with temperature may be put down to two conflicting factors: an increase in the proportion of 'holes' filled as the water temperature increases, that is, as more molecules become sufficiently mobile to fill them; and an increase in inter-molecular separation with temperature. The former tends to increase the density whilst the latter decreases it. From $0\ °C$ to $4\ °C$ the former is dominant; above $4\ °C$ the latter is dominant.

3.1.4 *Vapour pressure*

Water vapour is always present in the air. Near the ground, it contributes a partial pressure (see Section 2.1.1) of about 5 to 30 millibars. The fact that water vapour does exert a pressure can be demonstrated easily by putting a drop of water in the vacuum of a simple mercury barometer—as the water evaporates the level of the mercury falls. After complete evaporation of the drop, the level becomes steady and the fall measures the *vapour pressure*, *e*, of the water vapour in the space above the mercury which is now no longer a vacuum. The atmospheric pressure on the cistern now balances this vapour pressure in addition to the weight of unit cross-section of the mercury column. If more drops are added the mercury level falls further, but not indefinitely. In time a drop will persist, only partially evaporated, and the vapour pressure will then have reached a maximum known as the *saturated vapour pressure*, e_s. The vapour in the space above the mercury is now said to be *saturated* whereas before it was *unsaturated*. In general, if *e* is less than e_s a vapour is unsaturated and if *e* equals e_s it is saturated. In theory, it is possible to have *e* greater than e_s, when the vapour is said to be *supersaturated*, but in the atmosphere this occurs only under certain special conditions. Super-saturation of a vapour is impossible if it is in contact with a free surface of its own liquid because, when the two are brought together, the excess vapour condenses until it becomes just saturated. Hence, in defining e_s, the presence of such a surface should be indicated. Also, because of this, we can speak of the saturated vapour pressure *of the liquid*, meaning the maximum pressure of the vapour in the presence of the liquid. It follows from Dalton's Law (Section 2.1.1) that the value of *e* is independent of whether the water vapour is mixed with other atmospheric gases or not, so we can speak of the air being saturated with the vapour but at the same time meaning it is the *water vapour* which is saturated. It is quite common to hear references to the degree of saturation of the air rather than its constituent water vapour.

The numerical value of the saturated vapour pressure of a substance increases with the temperature, because the greater molecular separations and speeds in a liquid at higher temperatures allow the fastest molecules a greater chance to escape. Figure 20 shows the variation of e_s with temperature in the form of a graph known as an $e_s–T$ diagram; curve PW is the part of the diagram which refers to water. At 100 °C, e_s has a value equal to the normal atmospheric pressure and at this temperature boiling occurs. Boiling is a special type of evaporation where the phase change from liquid to vapour takes place within the liquid and not at its surface. Bubbles of vapour can form inside the liquid only when their pressure just exceeds the pressure exerted by the atmosphere on the liquid. When the atmospheric pressure is less than normal, boiling will occur at a lower temperature; thus, at the centre of a depression with a pressure of, say, 950 millibars, it boils at about 98·5 °C, and at the top of a 1000-metre high mountain where the pressure is about 900 millibars, it boils at about 97 °C. Values of e_s corresponding to ordinary temperatures are: at 0 °C, 6·1 millibars, at 10 °C, 12·3 millibars and at 20 °C, 23·4 millibars.

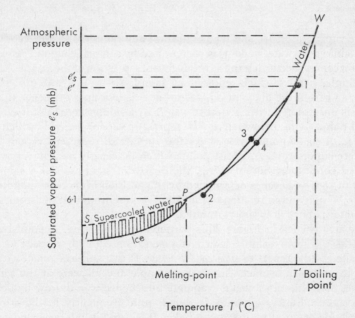

FIGURE 20. *Graph showing the variation with temperature of the saturated vapour pressure over flat surfaces of pure water and ice*

Values of e_s at different temperatures have also been measured for *supercooled* water, and its $e_s–T$ curve is found to be continuous with the curve for normal water—shown in Figure 20 as the broken curve PS.

The pressures of saturated water vapour over an *ice* surface have also been measured. The $e_s–T$ curve for ice, PI in Figure 20, meets the curve for water at 0 °C at point P, but elsewhere PI lies below PS. Thus, at any given temperature below 0 °C, the value of e_s over ice is less than the corresponding value of e_s over supercooled water, and the values of this difference are shown in Table I; it reaches a maximum at about −12 °C.

TABLE I. *Variations with temperature of saturated vapour pressure and relative humidity*

	Temperature (°C)										
	0	−2	−4	−6	−8	−10	−12	−14	−16	−18	−20
	Millibars										
e_s over supercooled water	6·11	5·27	4·55	3·91	3·35	2·86	2·44	2·08	1·76	1·49	1·25
e_s over ice	6·11	5·17	4·37	3·69	3·10	2·60	2·17	1·81	1·51	1·25	1·03
Difference	0·00	0·10	0·18	0·22	0·25	0·26	0·27	0·27	0·25	0·24	0·22
	Per cent										
Relative humidity w.r.t. ice when relative humidity w.r.t. water is 100 per cent	100	102	104	106	108	110	112	115	117	119	121
Relative humidity w.r.t. water when relative humidity w.r.t. ice is 100 per cent	100	98	96	94	93	91	89	87	86	84	82

The vapour pressure and temperature of any sample of water vapour (or of air containing the water vapour) is represented by a point lying on Figure 20, and this point can be used to determine the degree of saturation of the sample. Consider, for example, point 1 which represents a sample with $e = e'$ and $T = T'$. At T', $e_s = e_s'$, and from the diagram it is seen that e' is less than e_s', hence the vapour sample is unsaturated with respect to water. By a similar argument it can be seen that all points lying below curve PW represent vapour samples which are unsaturated. Also, all points lying above the curve PW represent vapour samples which are supersaturated, whilst points which lie exactly on PW represent just saturated conditions. Summarizing, the degree of saturation of a sample with respect to a given phase is determined by the position of its corresponding point on the e_s–T diagram in relation to the curve representing that phase: below, it is unsaturated; above, it is supersaturated.

The area shaded in Figure 20 is particularly interesting. A point lying in it represents a vapour sample which is *super*saturated with respect to *ice*, but *un*saturated with respect to *supercooled water*. If two surfaces, one of water and one of ice, were in contact with such a sample and all were at the same temperature, then the water would evaporate but the ice would grow by deposition of the vapour. Such a situation is found quite frequently in the atmosphere where clouds occur at temperatures below 0 °C, and it will be considered again in Sections 3.4.2 and 6.2.2.

3.1.5 *Some modified properties*

So far, we have studied the properties of water which is *pure* and *in bulk*. In the atmosphere water is often impure to some extent, with foreign substances dissolved in it; also, it is usually in the form of minute droplets or crystals. Both of these facts alter the properties of water to some extent.

Consider first the effects of a *solute*. Substances dissolved in water, such as salt, lower its saturated vapour pressure even when the temperature remains the same, and the extent of this lowering depends upon the concentration of the solute. Hence, the e_s–T curve for a salt solution lies below the curve for pure water, and the more concentrated the solution is the farther below the original does its curve lie. We can also see that a vapour sample which is saturated with respect to a salt-water surface will be unsaturated with respect to a pure-water surface, so a salt-water drop can persist without evaporation in an atmosphere which is somewhat unsaturated with respect to pure water.

Sea water contains about 35 parts of solute per 1000 parts of water, most of the solute being salt. It freezes at a lower temperature than fresh water (assuming no super cooling)—about −1.8 °C—and when it freezes *pure* ice crystallizes out. Also, the density of sea water is greater than that of fresh water (for example, at 0 °C it is about 1028 kg m^{-3}) so that when icebergs melt the resulting fresh water (icebergs form from *land* ice) floats on the surface of the sea, even though it may be colder than the sea beneath it. Sea water does not have a maximum density as does pure water.

When liquid water is present as a very small drop, its fastest molecules have a greater chance of leaving its highly curved surface than they have when the surface is flat; the smaller the drop, the greater the chance. Thus, at a given temperature a small drop has a larger value of e_s than a plane surface or, in other words, a small drop will persist without evaporation only if the surrounding vapour is supersaturated with respect to a plane surface. The smaller the drop, the greater the necessary degree of supersaturation. For drops with a radius greater than a few micrometres the effect is unimportant, but for smaller drops (for example, those which have just formed or those which are about to evaporate completely) the effect is an important factor in controlling a drop's size.

3.2 WATER VAPOUR IN THE ATMOSPHERE

3.2.1 *Humidity*

A measure of the amount of water vapour present in the air is known as the *humidity*. There are several ways of expressing it, one of which we have already met, namely, the vapour pressure. Four others are of importance.

(a) *Dew-point temperature*, defined as the temperature at which the air would become saturated if cooled at constant pressure. Figure 20 shows that for a given vapour pressure there is only one temperature at which the air will be just saturated, since the saturated vapour pressure is uniquely determined by the temperature. Hence, the dew-point temperature (commonly referred to as simply 'dew-point') is uniquely determined by the vapour pressure, and all samples of air having the same vapour pressure also have the same dew-point. If the air temperature of a sample is higher than its dew-point, the air is unsaturated; the closer are the two temperatures, the nearer is the air to saturation. If the air is cooled below its dew-point it tends to become supersaturated but, in practice, the water vapour in excess of that required for saturation condenses out. The dew-point of an air sample is independent of its temperature as long as it remains unsaturated, and the only way to alter the dew-point is to change its vapour pressure by evaporation or condensation. With air which becomes saturated only when cooled below 0 °C we have two possibilities:

it can become saturated with respect to either ice or supercooled water. The temperature at which air becomes saturated with respect to ice if cooled at constant pressure is known as the *frost-point*. Cooled to its frost-point, air is still unsaturated with respect to supercooled water (see Table I) and hence its dew-point has a value below that of the frost-point. The lower the frost-point, the greater the difference between them.

(b) *Humidity mixing ratio, r*, of moist air defined as the ratio of the mass of water vapour to the mass of dry air with which the vapour is associated. Since this ratio is small, r is usually expressed in g kg^{-1}. Common values are in the order of 5 to 30 g kg^{-1}. For each temperature there is a corresponding saturated humidity mixing ratio.

(c) *Relative humidity*, defined as the ratio e/e_s, where e_s is the saturated value at the air temperature. To a close approximation the corresponding ratio r/r_s of the humidity mixing ratios may be used. We may write:

$$\text{relative humidity} = \frac{\text{measured value of } e}{e_s \text{ at the air temperature}}$$

and, since the measured value of e is the same as the saturated value at the dew-point, then

$$\text{relative humidity} = \frac{\text{value of } e_s \text{ at dew-point}}{\text{value of } e_s \text{ at air temperature}}.$$

The relative humidity is usually expressed as a percentage. When the air is unsaturated, e is less than e_s and therefore the relative humidity is less than 100 per cent; for saturated air the relative humidity is 100 per cent. Unlike the dew-point, the relative humidity depends not only upon the vapour pressure but also upon the temperature.

(d) *Wet-bulb temperature*, defined as the lowest temperature to which an air sample can be cooled at constant pressure by evaporating water into it. If the air is originally saturated, evaporation cannot occur and there is no cooling. The wet-bulb temperature of saturated air is thus the same as the air temperature and, of course, the same as the dew-point. If the air is originally unsaturated, cooling will result from the evaporation and the wet-bulb temperature will be below the air temperature. When the air has been cooled to its wet-bulb temperature it will be saturated but its vapour pressure will be greater than originally, so the dew-point of the original air must be below its wet-bulb temperature. The relation between the three temperatures is thus always:

air > wet bulb > dew-point (unless all happen to be equal).

The difference 'air temperature minus wet-bulb temperature' is known as the *wet-bulb depression*; the larger its value the drier the air.

3.2.2 *Measurement of humidity*

The idea of wet-bulb temperature has arisen from the use of instruments known as *psychrometers* for measuring the humidity of the air. A psychrometer consists simply of two thermometers, one of which has its bulb permanently moist—the wet bulb. When air blows across the instrument, the wet-bulb thermometer reads lower than the dry bulb, and the depression gives a measure of the humidity, from which e can be calculated by using *Regnault's equation*:

$$e_s' - e = Ap\,(T - T')$$

where the symbols have the following meanings:

e = vapour pressure to be determined
e_s' = saturated vapour pressure at temperature, T'
p = pressure
T = dry-bulb temperature
T' = wet-bulb temperature
A = a constant.

On a given occasion, all the quantities in the equation are either known or can be measured, except e, which can thus be calculated. Having found e and knowing T, the remaining humidity variables (dew-point, humidity mixing ratio and relative humidity) can be calculated too.

The wet-bulb temperature is a *steady* temperature under a given set of conditions, which implies that there is no net loss of heat by the bulb. Since its temperature is below that of the air impinging upon it, heat must be flowing into the bulb by conduction, but its temperature remains steady because this flow is just balanced by the rate at which latent heat is used up while evaporation continues. Our own skins to some extent act like wet-bulb thermometers. Water on the skin's surface evaporates into the surrounding air, thus adding to the cooling already brought about by contact with the air. The effect is clearly most important on a windy day, which thus feels colder than a calm day even though the air temperature is the same. Since evaporation is slowest when the air already has a high relative humidity, we feel uncomfortable on a warm day with a high relative humidity, especially if the wind is light too, and we say the weather is 'muggy'. The effect of evaporation also explains why hot, dry, desert air is more comfortable than the relatively cooler but moist air of equatorial regions. In hot, dry climates, drinking water may be kept cool by storing it in a porous pot so that some of it is always slowly evaporating from the pot's outer surface.

Assmann has found that the constant, A, in Regnault's equation decreases as the speed of the draught over the wet-bulb thermometer increases so that, for a given humidity, the wet-bulb depression increases with the draught speed. However, this effect is important only with speeds below about seven knots; increasing the draught beyond this value has little further effect. The wet-bulb depression read from thermometers kept in a standard screen depends upon a draught speed of two to four knots maintained by the design of the screen. As long as the wind speed outside the screen at least equals this, the readings are reliable; if the wind is too light, the depression is too small and the calculated humidity is too large. An instrument which uses a forced draught of at least seven knots over the wet bulb is known as an *aspirated psychrometer*, and the corresponding temperature is the aspirated wet-bulb temperature.

When evaporation takes place from an ice film on the wet bulb, the resulting steady temperature is known as the *ice-bulb temperature*. If the water on the wet bulb remains supercooled when its temperature falls below 0 °C, then the reading will be lower than it would have been if the water had frozen and the thermometer had acted as an ice bulb. This is because supercooled water evaporates more readily than ice at the same temperatures—its saturated vapour pressure is always greater as we saw in Figure 20 and Table I.

In winter, the air is sometimes unsaturated with respect to supercooled water but supersaturated with respect to ice. A (supercooled) wet bulb would then read lower than the dry bulb in the normal way, but an ice bulb would read higher than the dry bulb because latent heat is liberated as water vapour is deposited on to it. The wet bulb can never be higher than the dry bulb but the ice bulb can be higher when these rather special conditions prevail.

Again in winter, on a day when the dry-bulb temperature is rising rapidly, an ice bulb will take a finite time to thaw, perhaps 15 minutes, and during this period it will record a constant temperature of 0 °C. This is, of course, a false reading; it is too low and will give a correspondingly low relative humidity.

Dew-point and vapour pressure show only slight *diurnal variations* but marked seasonal changes. Approximate values for Britain are: winter, 2 to 7 °C and 7 to 10 millibars; summer, 13 to 18 °C and 15 to 20 millibars. With clear skies the relative humidity shows a much more marked diurnal variation, with a minimum in the afternoon corresponding to the temperature maximum, and a maximum just after dawn. A normal diurnal range might be from 60 to 95 per cent. A relative humidity below 40 per cent is unusual, even in summer; only on very rare occasions has it fallen below 10 per cent in Britain.

Just as the temperature normally decreases with height so does the dew-point. A fall of dew-point with height is known as a positive *hydrolapse*. There is no limit to the size of the hydrolapse in the free atmosphere—a contrast to the maximum value of the environment lapse rate which cannot exceed the dry adiabatic lapse rate. (However, see Section 1.5.1.) On some days the hydrolapse may be very steep, even exceeding 50 °C km^{-1} in shallow layers. The relative humidity in the free atmosphere shows greater fluctuations than at the surface; values below 10 per cent are not infrequent in the middle troposphere and in the stratosphere.

When a parcel of air rises through the atmosphere, not only does its temperature fall but so does its dew-point, even whilst the parcel remains unsaturated. This is because the decrease of pressure which the parcel suffers as it rises affects in proportion the partial pressures of each constituent; hence the vapour pressure falls as the parcel rises, and since the dew-point is determined uniquely by the vapour pressure, it too must fall. The *adiabatic hydrolapse* for unsaturated air is 1·7 °C km^{-1}. It explains the need for specifying the condition of 'cooling at constant pressure' when defining the dew-point.

An instrument for measuring humidity is known as a *hygrometer*. A psychrometer is one type of hygrometer; the principles of some others in common use are:

(a) *Electrical absorption type.* The electrical resistance of a solution of a very soluble salt, such as lithium chloride, increases with the relative humidity, because as the humidity increases the solution becomes more dilute since water vapour can condense from the air on to a solution's surface even when the air is not saturated (see Section 3.1.5; also 3.3.2).

(b) *Condensation type.* A polished surface is cooled until condensation occurs—shown by the surface becoming dull. The temperature at which this takes place is the dew-point (or frost-point) which can therefore be measured directly. Brewer has developed a frost-point hygrometer to measure down to −80 °C for use by aircraft flying in the stratosphere.

(c) *Those measuring relative humidity.* Some substances increase in length as the relative humidity increases, examples being human hair and gold-beater's skin. The relative humidity is always with respect to water, even when the temperature is below 0 °C. These substances may also be used in radiosondes for measuring humidity in the upper atmosphere.

3.3 LIQUID WATER IN THE ATMOSPHERE

3.3.1 *Condensation processes*

When moist air is cooled sufficiently it becomes saturated. Further cooling tends to give supersaturation, but in practice the excess water vapour usually condenses, and it is in only special circumstances that the air remains supersaturated. The greater the degree of cooling, the greater is the amount of water vapour that condenses. This water may appear either on the ground as dew or hoar frost, or as a suspension of minute drops or crystals in the air—a fog or cloud.

There are three ways of cooling air to bring about condensation:

(a) *Adiabatic expansion*, when air rises to levels with progressively lower pressures. The rate of cooling equals the dry adiabatic lapse rate whilst the rising air is still unsaturated, but equals the saturated adiabatic lapse rate after saturation when, very roughly, one gram of water vapour condenses from each cubic metre of air for every kilometre of ascent. When the cork of a lemonade bottle is suddenly removed a temporary cloud formed by adiabatic expansion may be seen.

(b) *Contact with a cold object* whose temperature is below the dew-point of the air. This applies particularly to the ground which may be relatively cold either because it has lost heat by nocturnal long-wave radiation, or because warmer air is blowing over it after a cold spell. Everyday examples of this process are: walls becoming damp, or even running with water, when warm and moist air floods the country after cold weather, and also the 'steaming' of the insides of windows in winter, and of spectacles when entering a warm room.

(c) *Mixing* of two nearly saturated parcels of air having markedly different temperatures. Points 1 and 2 in Figure 20 represent two such parcels. If equal masses of each are thoroughly mixed the result would be represented by point 3 halfway between them, but since this represents a supersaturated state, the excess vapour condenses so that the vapour pressure of the mixture falls a little and its temperature rises (latent heat is released). The final result is represented by point 4. It can be seen that the amount of water condensed increases with the value of the temperature difference of the two original air parcels but, even so, the amount is always small compared with the vapour remaining uncondensed. If the original air samples are more unsaturated, condensation may not occur (contrast the visible condensation in one's breath on a cold day with its absence on a warm, dry day).

Of these processes, the first is the most important in the formation of clouds, and the second in the formation of fog, dew and hoar frost. The third process is normally found only along with the other two.

3.3.2 *Condensation nuclei*

Water vapour will not condense from perfectly clean air unless either it becomes very greatly supersaturated (relative humidities of several hundred per cent) or a free surface of water or ice is in contact with it. Now such high degrees of supersaturation are never found in the atmosphere and also, when condensation first occurs, free surfaces of water or ice are absent. Nevertheless, condensation is common, showing that it is still possible without these conditions, so we must conclude that it is the impurities in the air which promote condensation—specks of them act as centres upon which the condensation can start. These centres are known as *condensation nuclei* and their presence is essential to the initiation of condensation.

We can understand why this should be by considering what happens to the first few molecules of water vapour combining to form a new droplet.* The droplet is so minute that its extreme curvature requires a very high degree of supersaturation to allow it to persist (Section 3.1.5) and, in fact, the supersaturation is prohibitively high. If a solute is present in the water, however, the droplet can persist in an unsaturated atmosphere. These two effects conflict, so that minute droplets can form on soluble nuclei in the atmosphere even when it is unsaturated. As the droplet grows, becoming more dilute and less curved, the degree of unsaturation which it can tolerate without evaporation becomes smaller—for most droplets with radii greater than 5 μm a relative humidity of about 100 per cent is necessary for persistence (see, however, 'smog' in Section 4.4.1).

In recent years there has been much research into the nature and concentration of these nuclei, especially by Junge in Germany and by Twomey and Woodcock in the United States of America. It has been found convenient to divide the nuclei into three categories:

(a) Aitken nuclei with radii between 5×10^{-3} and 2×10^{-1} μm.
(b) Large nuclei with radii between 0·2 and 1 μm.
(c) Giant nuclei with radii greater than 1 μm.

Concentrations of Aitken nuclei vary greatly from as low as 10^6 per cubic metre over the oceans and over mountains up to 10^{12} per cubic metre near large industrial areas. Although the nuclei may be present in large numbers, mass concentrations are rarely as large as 10 μg per cubic metre and it is almost impossible to analyse their composition. By considering possible sources it seems most likely that they consist of combustion products such as ammonium sulphate and sulphuric acid. They are of little significance in the production of cloud droplets.

For large nuclei concentrations are usually in the range 10^6–10^9 per cubic metre, and for giant nuclei of the order of 10^6 per cubic metre. The composition of these nuclei is rather varied: the principal constituents in many cases are combustion products and soil particles (dust) with, particularly for giant nuclei, the addition of hygroscopic salts. Sea salt nuclei are formed over the oceans and their rate of formation varies with the wind speed.

The sizes and concentrations of these nuclei in the air control the sizes and concentrations of the droplets forming on them, and these in turn influence such

---

* The small drops present in clouds and fog, having radii less than 100 μm, are conveniently called 'droplets' to distinguish them from 'drops' of larger radii.

factors as the visibility (see Section 4.1) and the formation of precipitation (see Section 6.2.4). The small sizes of condensation nuclei would perhaps seem to make their identification a difficult task, but several interesting methods have been devised to do this. They include:

(a) Capturing the particles on fine threads (for example, spiders' webs) and examining them under a microscope, either optical or electron.

(b) Passing air containing the nuclei through a flame photometer, an instrument which enables the chemical composition of the particles to be found by examination of the characteristic wavelengths of the radiation emitted by them when heated to a high temperature.

(c) Capturing the particles on a chemically sensitive plate, each particle giving a permanent, coloured spot.

(d) Scattering from a volume containing many droplets or from a single particle.

3.4 ICE IN THE ATMOSPHERE

3.4.1 *The freezing of water*

Ice can form in two ways, either by the freezing of water or by the deposition of water vapour on to an already existing ice surface. Both occur in the atmosphere, and it is necessary to study them before we can understand such phenomena as the growth of precipitation.

In Section 3.1.1 we noted that water may be cooled below 0 °C without freezing, so it is of interest to see how this is possible. The explanation is similar to the one used for the initial condensation of water vapour to form a minute droplet (Section 3.3.2), namely, unless certain *freezing nuclei* are present, pure water will not freeze unless either it is greatly supercooled (to below the 'Schaefer point'—usually about −40 °C) or it comes into contact with an ice surface. In the absence of these nuclei or an already existing ice surface, freezing commences by the combination of a few water molecules to give an embryo ice particle whose extreme surface curvature allows its molecules to escape into the surrounding liquid more easily than from a plane ice surface. Consequently, until the temperature has fallen well below 0 °C, these particles are very short-lived. Below about −40 °C they can persist and water freezes spontaneously. The temperature is known as the 'Schaefer point'. It is to be expected that the chances of forming an embryo ice particle in some supercooled water increases with the volume of the water, hence larger drops are more likely to freeze spontaneously at higher temperatures. For most cloud droplets with radii between 1 and 100 μm the 'Schaefer point' is in the range −41 to −36 °C.

The presence of freezing nuclei induces supercooled water to freeze at temperatures above the 'Schaefer point'. Natural nuclei found in snow crystals have been identified as particles of certain clay minerals and other insoluble materials. By adding particles of these substances to artificial supercooled clouds it has been found that each substance has a 'threshold temperature' below which it is active in inducing freezing. For the most efficient nuclei this temperature is about −10 to −15 °C, so that water droplets in natural clouds are unlikely to freeze until cooled below about −10 °C. As the temperature decreases, so more nuclei become effective and the proportion of droplets which freeze increases. Many investigations of freezing nuclei have been made, for example, by Mason in Britain, Itoo, Kumai and Isono in Japan, and Schaefer and Hosler in the United States of America. There is still doubt as to how these nuclei cause freezing, but an essential property of a nucleus seems to be a capacity for adsorption of water molecules on its surface in such a way that the structure of the resulting film resembles that of ice. Once they have been involved in ice-crystal formation some ice nuclei become active at a higher temperature than initially and at humidities only slightly above saturation with respect to ice. This pre-activation may be destroyed by exposing the nuclei to air subsaturated with respect to ice or to temperatures above about -5 °C.

When water droplets freeze they largely retain their *spherical* shape. When bulk water freezes, *crystals* are formed having flat faces which are oriented at certain fixed, characteristic angles.

3.4.2 *Deposition of water vapour*

Crystals of ice formed by the freezing of *bulk* water have faces which are usually poorly developed; the best ice crystals appear when water *vapour* is deposited directly on to an ice surface. It has been found that for the initial formation of ice when moist air is cooled, saturation with respect to *water* must be reached (or at least approached) and not just saturation with respect to *ice*. This shows that condensation probably first takes the form of water droplets which later freeze.

Under conditions found in the atmosphere the normal form of an ice crystal is the regular *hexagonal prism* (Figure 21(*a*)) which may appear either flattened as a *plate*, because of growth parallel to the basal plane (Figure 21(*b*)), or elongated as a *needle*, because of growth parallel to the major axis (Figure 21(*c*)); intermediate forms are known as *columns*. Branching plates are common—both *stellar*, with six simple branches (Figure 21(*d*)), and *dendritic*, with six complex branches (Figure 21(*e*)). Only rarely are crystals perfectly symmetrical although the degree

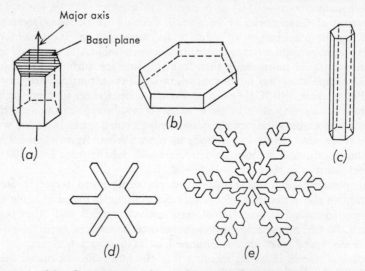

FIGURE 21. *Some common forms of ice crystals shown diagrammatically*
(a) Hexagonal prism. (b) Hexagonal plate. (c) Hexagonal needle. (d) Stellar crystal. (e) Dendritic crystal.

of symmetry is sometimes very large; most show hollows on their faces, and air bubbles inside. Combinations of the basic types are common, especially plates with branches at their corners, stellar crystals with plates at their tips, and columns with a plate at one or both ends (capped columns). Examples from this enormous variety of crystal shapes may be seen in snowflakes which are essentially loose aggregates of crystals. These crystals make a fascinating study, but their small size—usually 1 to 5 millimetres diameter—makes the use of a hand-lens an advantage.

Each crystal shape develops under certain special conditions present whilst it is growing, thus:

(a) *Temperature* controls the direction of growth. In the ranges 0 to −3 °C and −8 to −25 °C plates form; in the range −3 to −8 °C and below −25 °C needles and prisms form.

(b) *Supersaturation with respect to ice* controls the branching and detail. Growth at about ice saturation is slow and gives unbranched crystals with few imperfections, whereas growth at high ice supersaturation is rapid and gives skelton-like crystals with many hollows, and is also favourable to growth of branches—the best dendritic crystals grow in a supercooled water droplet cloud at about −12 to −16 °C, which has a large degree of supersaturation with respect to ice (Table I).

Crystals composed of several basic types show evidence of growth under changing conditions as may occur, for example, when they fall through layers of the atmosphere with differing temperatures and humidities, and such crystals have been grown artificially in the laboratory. Nakaya in Japan and Mason in Britain are among those who have investigated the problem.

Sharp edges and corners of crystals have higher saturation vapour pressures than flat faces (compare curved droplets with flat surfaces) so that on keeping, ice crystals slowly tend to lose their shape, becoming roughly spherical. This may be seen in old lying snow where the finest crystals have evaporated and the texture has become much coarser.

An interesting property of ice is its increase in *hardness* at low temperatures. At about −40 °C it is as hard as iron, so it is not surprising that sharp, wind-driven crystals in the Antarctic can be very painful when they collide with the skin.

3.5 CONDENSATION NEAR THE GROUND

3.5.1 *Dew*

In Section 1.5 we saw that, on a clear night, the ground temperature falls because of its continuous loss of heat by long-wave radiation. The air in contact with it is chilled, and when the temperature falls below the dew-point of the air, *dew* is deposited.

Not all parts of the ground cool at the same rate; for example, bare soil cools more slowly than vegetation because upward flow of heat within the soil by conduction from lower layers partly offsets the cooling by radiation, whereas vegetation may be considered as being thermally insulated from the ground by a 'blanket' of poorly conducting air among its leaves and so it cools more rapidly. At the same time the shielding effect of the vegetation prevents much loss of long-wave radiation directly from the soil beneath it, so that the ground temperature remains several degrees Celsius higher than that at the top of the canopy of vegetation above it. This shows us why dew forms preferably on vegetation rather than on bare soil and, further, on the tops of the plants rather than lower down.

This theory of dew formation by condensation from the air was put forward by Wells in 1818. Later work by Monteith and by Long has added to our understanding of the process. Consider a grass surface. When the temperature of the leaves falls below the dew-point of the air next to them, water vapour condenses on the the numerous nuclei on their surfaces. While the temperature continues to fall, the dew deposit increases. There are two sources for this water vapour: the air above, and the ground, from which evaporation continues as long as its temperature remains above the dew-point of the air (which it usually does throughout the night). If the wind remains light, below about 1 knot at a height of 2 metres, the transfer of water vapour downwards by turbulence is small and most of the dew originates from vapour derived from the soil beneath. With rather stronger winds, 1 to 6 knots at 2 metres, the downward transfer is predominant. In the first situation the dewfall is only light, usually less than 0·1 millimetres during a summer night; in the second it may be heavy, say, 0·1 to 0·15 millimetres. If the wind is stronger than about 6 knots at 2 metres, the air is prevented from cooling below its dew-point because the increased turbulence causes excessive mixing with warmer, drier air aloft.

Conditions favourable for dew formation are thus seen to be:

(a) Clear night sky (to allow maximum loss of long-wave radiation).
(b) Moist air at sunset (relative humidity preferably greater than 75 per cent).
(c) Wind of about 1 to 6 knots at a height of 2 metres.

The age-old question of whether dew 'rises' or 'falls' is now answered: it does either or both, depending upon the conditions.

Note should be made of the fact that not all the water present on leaves after a clear night is necessarily dew. There are two other sources:

(a) *Fog-water*, resulting from the collection of water droplets on the leaves as a fog blows past them. Water gathered by trees in a fog may fall to the ground as *fog-drip*.
(b) *Guttation-water*, exuded from the leaves. The relative warmth of the plant's roots compared with its leaves allows the upward flow of sap to continue at night. Water passing through the leaf pores collects because evaporation is impossible into the surrounding saturated air. Guttation drops are sparse but large (greater than two millimetres diameter) whereas most dewdrops are more numerous and smaller (less than one millimetre).

3.5.2 *Hoar frost*

If the condensation of water vapour does not begin until the temperature is below 0 °C then the initial water droplets soon freeze and deposition of the vapour continues with the formation of soft, white ice crystals known as *hoar frost*. The crystals assume the forms described in Section 3.4.2 but are usually distorted because of the unsymmetrical distribution of vapour around their growing parts. A night's deposit exceeding the equivalent of 0·1 millimetre of rainfall is unlikely.

On some nights, dew which has been deposited at temperatures greater than 0 °C may be seen to cool below 0 °C without freezing. Such *supercooled dew* usually freezes if its temperature falls below about −3 to −5 °C giving a deposit known as *frozen dew, white dew* or *silver frost*. Both frozen dew and hoar frost may coexist later in the night. Water formed by the melting of hoar frost as morning temperatures rise should not be confused with dew. Both hoar frost and dew are almost always formed by radiative cooling of the ground, but deposits are sometimes found as the result of advection of moist air over a cold land surface. Both processes together can cause copious dewfall on a clear summer night, especially near a windward coast.

The fern-like patterns of *window frost* form when a crystal grows at the expense of supercooled droplets on the glass. During cold weather these droplets appear on the inside of a window, and the atmosphere in their vicinity is saturated with respect to water so that if one of them freezes it is surrounded by air which is supersaturated with respect to ice (see Section 3.1.4 and Table I) and it grows into a branched crystal. The branches slowly creep over the glass and the droplets just near them evaporate. Careful inspection shows a narrow, clear gap between a crystal and its neighbouring droplets.

Other deposits of ice on the ground are:

(a) *Ground ice*, formed by the freezing of lying water, by the re-freezing of partly melted snow, or by the compaction of snow on roads by traffic.

(b) *Glazed frost* (see Section 6.3.5).

(c) *Rime* (see Section 4.2.2).

BIBLIOGRAPHY

BANNON, J.K. and STEELE, L.P.; 1960. Average water-vapour content of the air. *Geophys Mem*, **13**, No. 102.

FOLLAND, C.K.; 1975. The use of the lithium chloride hydrometer (dew-cell) to measure dew-point. *Met Mag*, **104**, pp. 52–56.

HAY, R.F.M.; 1956. Ice accumulation upon trawlers in northern waters. *Met Mag*, **85**, 225–229.

MONTEITH, J.L.; 1957. Dew. *Q J R Met Soc*, **83**, pp. 322–341.

MOSSOP, S.C.; 1970. Concentrations of ice crystals in clouds. *Bull Amer Met Soc*, **51**, pp. 474–479.

OHTAKE, T. and JAYAWEERA, K.O.L.F.; 1972. Ice crystal displays from power plants. *Weather*, **27**, pp. 271–277.

PARKER, G. and HARRISON, A.A.; 1967. Freezing drizzle in south-east England on 20 January 1966. *Met Mag*, **96**, pp. 108–112.

PEDGLEY, D.E.; 1969. Snow and glaze on Christmas Eve 1968. *Weather*, **24**, pp. 480–485.

CHAPTER 4

VISIBILITY

4.1 INTRODUCTION

4.1.1 *Some definitions*

The term *visibility* may be defined as the greatest distance at which an object of specified characteristics can be seen and identified by an observer with normal sight under normal conditions of daylight illumination. It is measured by selecting a number of objects, such as church towers or hills, whose positions are known, and then determining at the time of observation which object can be seen whilst the one at the next greater distance cannot be seen. On any given occasion the visibility depends not only upon the opacity of the air, resulting from particles in suspension, but also upon such factors as the sensitivity of the observer's eye and the position of the object relative to the sun or moon. Here we will be concerned only with the meteorological factors governing the opacity of the air.

In the absence of any particles in suspension, visibility through the atmosphere is about 240 kilometres. It is probably true to say that in most places such an extreme is never reached; there are always some particles present. These are of three types:

(a) Minute particles of smoke, dust or water, so small that they settle only at extremely low wind speeds. If the particles are of smoke or dust they produce a *haze*; if of water (or ice) they produce a *mist* or *fog*.

(b) Coarse particles, large enough to be kept in suspension only by turbulence accompanying strong winds. Examples are: drifting sand or dust, and drifting snow (a 'blizzard'). In general, the stronger the wind the more seriously is the visibility affected by drifting. Duststorms are not unknown even in Britain; they can occur, for example, in the Fen country when strong winds blow across fallow fields after a dry spell. For an example, see *Meteorological Magazine*, **86**, 1957, p. 23.

(c) Precipitation. The worst visibility is associated with large numbers of small particles. Thick drizzle and heavy, fine snow can reduce the visibility to a few hundred metres.

It is important to distinguish carefully between fog, mist and haze.

(a) *Fog*: Obscurity in the surface layers of the atmosphere which is caused by a suspension of water droplets, with or without smoke particles, and which is defined by international agreement as being associated with visibility of less than 1 kilometre. In general poor visibility does not constitute a hazard to surface transport until it falls below about 200 metres. It is therefore British practice to use this term fog in forecasts for the general public to relate to visibility less than 200 yards (180 metres).

(b) *Mist*: A state of atmospheric obscurity produced by suspended microscopic water droplets or wet hygroscopic particles. The visibility is equal to or exceeds 1 kilometre; the corresponding relative humidity is greater than 95 per cent.

(c) *Haze*: A suspension in the air of extremely small dry particles invisible to the naked eye and sufficiently numerous to give the air an opalescent appearance. There is no upper or lower limit to the horizontal visibility in the presence of which haze may be reported, although it is very rare for haze to reduce visibility to less than 1 kilometre. Relative humidity is less than 95 per cent. In most cases the particles composing haze are small enough to cause differential scattering of sunlight and to contribute for example to sunrise and sunset colours.

In Britain, most mists and fogs form in air which has been cooled by contact with a cold ground. Whilst the cooling is in progress the relative humidity increases, and a stage is reached when water vapour starts to condense on the nuclei in the air. The relative humidity is then about 95 per cent and the droplets formed are small and sparse—a mist is formed. With progressive cooling, the relative humidity increases and the droplets grow in size and number so that the visibility decreases and may eventually fall below 1 kilometre to give a fog. From this we see that a fog is always preceded by a mist stage; similarly, mist is present after the dispersal of fog. On either occasion the mist stage may be extremely short lived. Note that if a fog has already formed elsewhere and is then brought in by the wind no mist stage may be seen.

There are two weather situations in which the ground is cooler than the air in contact with it: first, when the ground loses heat at night in the form of long-wave radiation, and second, when warm air flows from a warm region to cover a cold surface by advection. Both can produce fogs, in the first situation we have a *radiation fog*, and in the second an *advection fog*.

4.2 RADIATION FOG

4.2.1 *Formation and properties*

We are all familiar with the type of fog present early on some mornings after a clear, quiet night—a fog which disperses during the morning and which, in popular weather-lore, is supposed to herald another day of good weather. There is some truth in this supposition as can be seen by studying the conditions favourable to its formation. In Section 1.5.1 we saw how the air temperature near the ground changes during a clear night. If the air is moist enough originally, the temperature may well fall below the dew-point and dew is deposited (Section 3.5.1). For fog to form the air well above the ground must also be cooled. There are two ways of doing this:

(a) By turbulence, when the wind blows over rough ground. The air cooled by conduction to the ground mixes with warmer air aloft, and the depth through which this takes place increases with the wind speed—it is usually at least several hundred metres deep. Lower dew-points are also spread upwards in this way because the air near the ground has been depleted of some of its water vapour by dew deposition.

(b) By radiation from one layer of air to another or from the droplets in suspension. This is a complex process but the final result is a little cooling up to about 1000 metres, the temperature fall being greatest near the ground.

The first method seems capable of going as far as a saturated stage only; it is the second which goes further, giving condensation.

On a night during which radiation fog subsequently forms, careful observation shows that not only does the temperature at screen-level fall but so does the dew-point. This is to be expected since dew is being deposited, so removing water vapour from the air. If the air is to become saturated the temperature must fall more quickly than the dew-point, but observations show that it is still a slow process—it may take several hours for the relative humidity to rise from 95 to 100 per cent. The temperature at which fog eventually forms—the *fog-point*—may be several degrees Celsius below the original dew-point. Fog formation is usually rapid in air that is free from smoke pollution, and the visibility can fall from 3 kilometres to 200 metres in perhaps 10 minutes. This rapid formation can sometimes be associated with a relatively sudden, although small, fall in temperature, and it seems likely that this results from cooling of the air by the second method, that is, long-wave radiation from the water droplets themselves. In a mist, droplets are few so radiational cooling is slight, but as they grow in size and number the loss of heat by radiation increases, causing further cooling and growth. The process accelerates, ending with rapid fog formation. In a clean water fog visibility seldom falls below 50 metres.

Fogs vary greatly in depth: most are in the range 15 to 100 metres, but on extreme occasions the top may be at 200 or 300 metres.

Some interesting temperature changes take place in a fog after it has formed. If it is deep and dense enough, the fog droplets prevent the loss of most of the long-wave radiation from the ground in the same way that a cloud layer does (Section 1.5.2), so the ground temperature rises, often quickly, as a result of heat conduction from lower levels in the ground, until it is about the same as the air temperature. This rise of the ground temperature can be followed by noting the temperatures recorded by a grass-level thermometer. Droplets near the fog's top continue to radiate upwards, so the upper part of the fog continues to cool, thus causing the lapse rate in the fog to increase. Cooling by long-wave radiation continues by day and night, and in winter can predominate over the gain of heat from the weak insolation, only a small fraction of which reaches the ground. The final result of these processes is a replacement of the original pronounced inversion with its base on the ground by a layer with its lapse rate equal to the saturated adiabatic lapse rate and topped by a strong inversion. The fog's top lies near the base of this inversion at a height of several decametres. High ground may project upwards through the fog into the clear, warm, dry air above. Radiation fog is notoriously patchy, forming usually in certain favoured localities, but occasionally it can be widespread.

From the above discussion of its formation we can see that the conditions favourable to the development of radiation fog are:

(a) A clear night sky to allow maximum loss of radiation, preferably long winter nights.

(b) Moist air at sunset, most often found in autumn and winter, especially after rain or near open water.

(c) A light wind (approximately one to four knots) to give sufficient turbulence to spread cooling upwards. Anticyclones are favourable; so are sheltered valleys.

These conditions are the same as those favourable to dew (see Section 3.5.1) and, in fact, radiation fog is likely to be preceded by dew. However, dew is not always followed by fog and this may be explained nearly always by a lack of sufficient time which will allow the fog-forming processes to be completed before fog-dispersing processes become dominant. Radiation fog disperses by either an increase in wind speed, resulting in a mixing of the fog-laden air with warm, dry air aloft, or more commonly by insolation warming the ground and this in turn warming the foggy air in contact with it. Either way may result in the temporary formation of cloud with a low base, known as 'lifted fog' (see Section 5.3.2). It should be noted that radiation fog does not form over a large water surface because its diurnal temperature change is far too small.

4.2.2 *Wet fog and rime fog*

Fog drifting across vegetation deposits fog-water on to the leaves and the amount may become considerable; in some places this water adds appreciably to the rainfall. A fog depositing water is said to be a *wet fog*; its droplets are usually larger than normal.

Fogs can exist at temperatures below 0 °C, the droplets then being super-cooled. Temperatures below −10 °C have been recorded in British radiation fogs but they are unusual. When *supercooled fog* drifts past obstacles, such as vegetation or fences, some of the droplets are caught and they freeze almost instantaneously to form a white, opaque deposit comprising numerous frozen droplets with enclosed air-spaces. This deposit is known as *rime*; it is similar to hoar frost in that it is fragile and has a low density but is distinguished from it by its non-crystalline character and by its building on the upwind sides of objects. In a persistent, supercooled *rime fog* the deposit may grow at the rate of 1 centimetre per day. Dense supercooled clouds enveloping mountains can cause rime to grow at much greater rates.

4.2.3. *Ground fog*

On some evenings when the ground is very wet and the lowest layers of the atmosphere are nearly saturated, say following afternoon rain, a shallow layer of fog develops, often around sunset. It is so shallow, 1 to 2 metres deep, that the upper parts of bushes and cattle may be seen standing clearly above it, and for this reason it is known as *ground fog*. The horizontal visibility at eye-level may be several kilometres but within the fog it is only a few hundred metres. It is often patchy and seldom lasts more than one hour. Its rapid formation, usually when there is almost no wind, indicates that the air is not cooled by contact with the ground but by direct radiation, probably from small droplets suspended within it.

4.3 ADVECTION FOG

4.3.1 *Formation and properties*

Whereas radiation fog forms only inland, over the ocean and on windward coasts another type of fog is common. This is found when moist air flows from a warm sea surface across one which is colder—one whose temperature is below the dew-point of the air—so this fog, formed by advection, is known as *advection fog*. The wind which necessarily accompanies advection is the agent for spreading the cool lowest layers of air through a greater depth. The stronger the wind the greater the depth of the layer in which cooling takes place, and hence the greater the cooling necessary to produce a fog. If the wind is too strong or the cooling too small or both, only low cloud is formed (see Section 5.3.2) or perhaps there is no condensation at all.

In Britain, there are two principal examples of this type of fog:

(a) With moist south-west winds flowing across a progressively colder sea surface as Britain is approached, extensive fog develops (*sea fog*) and it invades all exposed coasts (*coast fog*) in the west and south. If the wind is greater than about 10 to 15 knots, it is usually replaced by low cloud with a base of a hundred metres or so.

(b) With east winds flowing across the cold North Sea in summer. Over the continent the air is very warm, but on crossing the sea it is cooled and moistened from below so that by the time it reaches our east coast it may be fog-laden.

Over the sea advection fog has little or no diurnal variation, but as it spreads inland sunshine usually disperses it, especially in the summer; it returns again at night by spreading in from the sea. Its reformation is often aided by radiational cooling of the land. On windward coasts it can persist all day with cool, clammy conditions, whereas perhaps ten miles inland the day is bright, warm and dry. Advection fogs can also be found when moist air flows over a land surface after a cold spell, particularly when the surface is covered by thawing snow which effectively keeps the ground temperature at 0 °C.

4.4 OTHER TYPES OF FOG

4.4.1 *Smoke fog*

Most types of fog in Britain form either by radiative or advective processes. There are others, however, which although less common are still of interest. Near large urban and industrial areas the atmosphere usually contains sufficient smoke to add appreciably to the concentration of small condensation nuclei. When fog forms in this air the droplets are very small and are present in huge numbers. Condensation occurs when the relative humidity is less than 100 per cent so that a noticeable reduction of visibility is produced even when the air is far from saturated. It is not unusual to find a visibility below 1 kilometre when the relative humidity is only 90 to 95 per cent and sometimes even less. *Smoke fog* (smog), therefore, forms earlier than would a purer water fog under the same conditions. Whilst cooling continues the fog thickens, and the visibility may eventually fall below five metres in a 'pea-souper'. The persistence of the droplets in unsaturated air also ensures that the visibility remains below 1 kilometre for a longer period when the fog is dispersing. The very numerous and effective nuclei in smoke fog thus explain its well-known features—persistency and denseness.

In the middle of a city this process may be modified by the conglomeration of houses, factories, office blocks and surfaced roads. As cooling occurs generally, these provide a source of heat and air temperature falls less than in the surrounding suburbs and open country. The effect is noticeable not only in minimum temperatures recorded (the difference between central London and the green belt may be as much as 8 °C) but also in the frequencies of thick and dense fog. A comparison between central London (Kingsway) and outer London (represented by Heathrow and Kew) showed that thick or dense fog occurred about three times as frequently in outer London. Further out still the fog is cleaner of course, and so less dense again. Most smoke fogs are modified radiation fogs and show a normal diurnal variation, but in winter they can persist for several days in succession. The inversion which develops at the top of a layer of radiation fog traps the smoke beneath it and also prevents mixing between the foggy air and the unsaturated air above because of the increased stability. Any smoke put into the fog thus gradually increases in concentration while the fog persists.

4.4.2 *Steam fog*

When cold air passes over warm water the vapour escaping from the water surface is cooled immediately above it and condenses again in the form of steam, giving a *steam fog*. It forms in just the same way as steam over a hot bath, or over a wet road which is rapidly heated by the sun after a shower, but is best seen on a cold winter morning over ponds, rivers and canals where it sometimes accumulates into extensive patches. In arctic regions, when cold air from the interiors of the continents blows across the relatively warm sea, steam fog may be widespread and dense; it is then known as *arctic sea smoke*, or *frost smoke*. Because the air is usually so much colder than the sea (usually at least 10 °C) the lapse rate near the water is very steep, that is, the lowest layers are very unstable, so the fog appears as innumerable convective swirls each of which soon evaporates into the drier air above, giving a top at about 10 metres. Over land it may accumulate below a night-time inversion.

4.4.3 *Hill fog*

This is a term generally used of low cloud which envelops high ground. The production of saturation and condensation by forced uplift of moist air is not necessarily implied in the use of this term as in the case of upslope fog.

4.4.4 *Ice fog*

At very low temperatures some of the water droplets in a fog may freeze and grow at the expense of the remaining drops which evaporate, giving an all-ice fog composed of minute crystals which scintillate in the sun and give to this *ice fog* its popular name of 'diamond dust'. Refraction of light through the crystals gives optical phenomena (see Section 8.1). Ice fogs are very rare in Britain but more common over the interiors of northern continents in winter and over Antarctica. With temperatures below about −30 °C nearly all fogs are of this type. In the coldest weather, towns, and even herds of caribou in North America, have around them banks of a type of artificial ice fog, produced by the spontaneous freezing of droplets in steam at temperatures below about −40 °C (see Section 3.4.1). This steam is derived from such sources as central heating, exhaust from cars and aircraft, and the breath of animals. Visibility is often much poorer than in natural ice fogs which usually contain fewer ice particles in suspension.

BIBLIOGRAPHY

BINDON, G.J.; 1962. Fog at Liverpool Airport. *Met Mag*, **91**, pp. 162–167.

BRIGGS, J.; 1969. Visibility variations at London/Heathrow Airport. *Met Mag*, **98**, pp. 135–138.

JENKINS, I.; 1971. Decrease in the frequency of fog in central London. *Met Mag*, **100**, pp. 317–322.

KELLY, T.; 1971. Thick and dense fog at London/Heathrow Airport and Kingsway/Holborn during the two decades 1950–59 and 1960–69. *Met Mag*, **100**, pp. 257–267.

ODDIE, G.J.W.; 1968. The transmissometer. *Weather*, **23**, pp. 446–455.

SAUNDERS, W.E.; 1971. Visibility deteriorations during winter mornings. *Met Mag*, **100**, pp. 149–155.

SAUNDERS, W.E.; 1973. Radiation fog and stratus formation and fog clearance in terms of geostrophic wind—some applications of wind measurements on a high mast. *Met Mag*, **102**, pp. 140–146.

CHAPTER 5

CLOUDS

5.1 CLASSIFICATION

5.1.1 *Introduction*

A *cloud* may be defined as a visible aggregate of minute particles of water or ice, or both, in the free air. From even a brief observation of the sky two fundamental characteristics of clouds become apparent: first is their infinite variety of form, and second their continual changes in appearance. It is the purpose of this chapter to describe and classify the forms most commonly observed and then to explain their appearance and evolution in terms of the physical processes taking place in the atmosphere.

In 1803 Luke Howard proposed a simple scheme of classification. He distinguished three principal cloud forms:

(a) *Stratus* cloud, lying in a level sheet.

(b) *Cumulus* cloud, having flat bases and rounded tops, and being generally lumpy in appearance.

(c) *Cirrus* cloud, having a fibrous or feathery appearance.

As our knowledge of clouds grew, this system proved inadequate and more careful and detailed observations enabled a far more comprehensive system to be devised. To attain a degree of world-wide uniformity in the naming and identification of clouds the International Meteorological Committee published a cloud atlas in 1895. Since then several revised editions have appeared, the latest being in 1975 when the World Meteorological Organization, a specialized agency of the United Nations, published the *International cloud atlas*. It gives a detailed account of the present system of classification, illustrated by a large number of carefully selected photographs, and is a valuable reference book which should be consulted whenever doubt arises over cloud classification.

5.1.2 *International system of classification*

The international system of cloud classification resembles the systems used in the botanical and zoological sciences. The different clouds are given descriptive names which depend mainly upon appearance, but also sometimes upon processes of formation as seen by an observer. Despite their infinite variety of forms it is possible to define 10 basic types of world-wide occurrence. These are taken as 'genera'. Most genera possess several 'species' and many of these, in turn, occur as a number of 'varieties', sometimes accompanied by supplementary features and accessory clouds. Descriptions of the genera are given below, whilst descriptions of the more important species and varieties will be included in later sections. They are taken from the 1975 edition of the *International cloud atlas*, and it is assumed that the clouds are observed from the ground on a clear day and under normal illumination by sun or moon.

(a) *Cirrus (Ci)*. Detached clouds in the form of white, delicate filaments,

65

or white or mostly white patches or narrow bands. These clouds have a fibrous (hair-like) appearance or a silky sheen or both.

(b) *Cirrocumulus (Cc)*. Thin, white patch, sheet or layer of cloud without shading, composed of very small elements in the form of grains, ripples, etc., merged or separate, and more or less regularly arranged; most of the elements have an apparent width of less than one degree (approximately the width of the little finger at arm's length).

(c) *Cirrostratus (Cs)*. Transparent, whitish cloud veil of fibrous or smooth appearance, totally or partly covering the sky, and generally producing halo phenomena (see Section 8.1).

(d) *Altocumulus (Ac)*. White or grey, or both white and grey, patch, sheet or layer of cloud, generally with shading, composed of laminae, rounded masses, rolls, etc., which are sometimes partly fibrous or diffuse, and which may or may not be merged; most of the regularly arranged small elements usually have an apparent width of between one and five degrees (approximately the width of three fingers at arm's length).

(e) *Altostratus (As)*. Greyish or bluish cloud sheet or layer of striated, fibrous or uniform appearance, totally or partly covering the sky, and having parts thin enough to reveal the sun at least vaguely, as through ground glass. Altostratus does not show halo phenomena.

(f) *Nimbostratus (Ns)*. Grey cloud layer, often dark, the appearance of which is rendered diffuse by more or less continually falling rain or snow which in most cases reaches the ground. It is thick enough throughout to blot out the sun. Low, ragged clouds frequently occur below the layer with which they may or may not merge.

(g) *Stratocumulus (Sc)*. Grey or whitish, or both grey and whitish, patch, sheet or layer of cloud which almost always has dark parts, composed of tessellations, rounded masses, rolls, etc., which are non-fibrous (except for virga—see Section 5.4.4) and may or may not be merged; most of the regularly arranged small elements have an apparent width of more than five degrees.

(h) *Stratus (St)*. Generally grey cloud layer with a fairly uniform base, which may give drizzle, ice prisms or snow grains (see Section 6.1.1). When the sun is visible through the cloud its outline is clearly discernible. Stratus does not produce halo phenomena (except possibly at very low temperatures). Sometimes stratus appears in the form of ragged patches.

(j) *Cumulus (Cu)*. Detached clouds, generally dense and with sharp outlines, developing vertically in the form of rising mounds, domes or towers, of which the bulging upper part often resembles a cauliflower. The sunlit parts of these clouds are mostly brilliant white; their bases are relatively dark and nearly horizontal. Sometimes cumulus is ragged.

(k) *Cumulonimbus (Cb)*. Heavy and dense cloud, with a considerable vertical extent, in the form of a mountain or huge towers. At least part of its upper portion is usually smooth, or fibrous or striated, and nearly always flattened; this part often spreads out in the shape of an anvil or vast plume. Under the base of this cloud, which is often very dark, there are frequently low ragged clouds either merged with it or not, and precipitation, sometimes in the form of virga.

5.1.3 *Other methods of classification*

Nearly all clouds occur in the troposphere between extreme heights of sea level and approximately 18 kilometres; the exceptions are considered in Section 5.7. A long-established sub-division of the troposphere into three layers is still used when describing the heights at which the bases of clouds occur—low, medium and high. The approximate height-ranges vary with latitude and are indicated in Table II.

TABLE II. *Approximate height-ranges at which bases of clouds are found*

Level	Height-ranges in polar regions (km)	Height-ranges in temperate regions (including Britain) (km)	Height-ranges in tropical regions (km)
High	3–8	5–13	5–18
Medium	2–4	2–7	2–8
Low		From earth's surface to 2 kilometres	

To some extent the clouds in each of these three layers are distinctive. Thus *Ci*, *Cc* and *Cs* are normally high clouds, *Ac*, *As* and *Ns* medium clouds, *St*, *Sc*, *Cu* and *Cb* low clouds, but not invariably so. These relationships can be largely put down to the ranges of temperature usually found in each layer. The appearance of a cloud depends to a great extent upon the nature of its constituent particles and also upon the concentration of water present, both of which are largely controlled by the temperature. We may illustrate this idea by considering *Ci* and *Cs*, both of which are composed of ice crystals sparsely distributed, thus allowing much of the sunlight to pass between them, so it is not surprising that these clouds occur at temperatures well below 0 °C, that is, at high levels. Again, *St* and *Sc* are composed usually of water droplets present in a sufficient concentration to obscure the sun, so one is most likely to find them at temperatures above 0 °C, that is, usually at low levels.

Table III gives common heights of cloud bases for each genus found over Britain.

TABLE III. *Approximate heights and temperatures of cloud bases commonly found over Britain*

Genus	Height of base (km)	Temperature likely to be found at base level (°C)
Ci *Cc* *Cs*	5–13	−20 to −60
Ac *As*	2–7	+10 to −30
Ns	1–3	+10 to −15
Sc	½–2	+15 to −5
St	surface–½	+20 to −5
Cu	½–2	+15 to −5
Cb	½–2	+15 to −5

Luke Howard's original classification can still be used in a broad sense to separate clouds into three fundamental forms:

(a) layered, or *stratiform*, clouds;
(b) heaped, or *cumuliform*, clouds;
(c) fibrous, or *cirriform*, clouds.

Although these terms are purely descriptive, each in fact has been found to correspond to a particular condition of formation, and similar relationships between appearance and specific methods of formation also apply to details of structure, many of which will be dealt with in subsequent sections.

5.2 FORMATION AND DISPERSAL OF CLOUDS— A GENERAL SURVEY

5.2.1 *Formation of clouds*

Clouds are continuously changing. Sometimes this can be seen very clearly, as with cumulus for example; with other clouds, such as cirrus, the changes take place very slowly. A cloud must not be looked upon as an entity drifting with the wind across the sky, but rather as a visible manifestation of physical processes taking place in the atmosphere, some of which tend to form the cloud while others disperse it. When formative processes predominate the cloud grows.

Two important facts which decide the appearance of a cloud are:

(a) stability of the atmosphere in which it forms,
(b) growth of precipitation within it.

Consider the first point. The stability determines the nature of the vertical motion in the atmosphere, for if it is stable the air must be forced upwards and the resulting motion is smooth, slow and usually widespread, so that the clouds formed by the adiabatic expansion accompanying the ascent are layered and widespread and they form slowly—they are stratiform. If the atmosphere is unstable, portions of the air may rise spontaneously and the resulting motion is irregular, rapid and usually localized, so that the clouds formed are heaped and scattered and they form quickly—they are cumuliform. Hence, we associate *stratiform* clouds with a *stable* atmosphere, and *cumuliform* clouds with an *unstable* atmosphere.

Next, consider the growth of precipitation within a cloud. When this occurs, particles appear which are much larger than the cloud particles originally present. Their sizes vary, but always they occur in a great variety, and since the largest fall fastest, gravity sorts them in such a way that soon after formation the largest appear below and the smallest above. The streamers of precipitation, or *fall-streaks*, so formed, take on a fibrous structure if they are small and detached, as with cirrus, but if they are widespread and fall from an extensive cloud they give to it a diffuse base with little or no visible structure, as with nimbostratus (see Plate I). *Cirriform* clouds are thus identified with isolated streamers of precipitation.

For a cloud to form, a part of the atmosphere must be cooled below its dew-point. When this occurs the excess water vapour, other than that required to

CLOUD
PLATES

D. E. Pedgley

Stratus and nimbostratus

The sky is completely covered with a thick dark layer of *Ns*. Below is a patch of low *St fra* formed when air near the surface is forced to rise over the hill in the distance. Its low base, about 100 metres, shows that this surface air is already nearly saturated before lifting.

PLATE I

D. E. Pedgley

Altocumulus

The very extensive layer of *Ac* (species *str*) is rather dense (variety *op*) and shows roughly parallel bands (variety *un*) which seem to radiate from a point on the horizon —an illusion caused by perspective.

PLATE II

Cirrocumulus, cirrostratus and cumulus

The sky is almost covered by a delicate veil of *Cs* and *Cc* at about 10 000 metres. The *Cc* appears as a thin layer (species *str*) with delicate ripples (variety *un*). Below are some rather ragged fragments of *Cu* (species *fra*) based at about 1000 metres.

PLATE III

D. E. Pedgley

Altocumulus

The series of thin almond-shaped clouds (species *len*) at about 4000 metres were formed orographically as the wind blew over the mountains in the Isle of Man. Some of the clouds are smooth, others are broken into the variety *pe*.

PLATE V

D. E. Pedgley

Stratocumulus

The extensive thin layer of *Sc* (species *str*) at about 2000 metres is broken into innumerable patches touching each other, between which some clear sky is visible (variety *pe*) giving a characteristic 'crocodile-skin' pattern. *Ac* also commonly occurs in a similar form.

PLATE IV

Cirrostratus and cumulus

The thin veil of *Cs*, probably higher than 10 000 metres, shows little detail. Below are many small *Cu* clouds with bases at about 1200 metres, some of species *hum* with flat bases and tops and others of species *fra*, ragged with no definite form. This combination of clouds is typical of the outer edges of an extensive area of multi-layered cloud and widespread rain, with shallow convection near the ground.

PLATE VI

Altostratus and cumulus

The higher layer of cloud is *As*, typically diffuse and grey, through which the sun shines dimly as through ground glass (a 'watery sky'). It is often found to follow the *Cs* such as is shown in Plate VI when a widespread rain area is approaching the observer. It is a deeper cloud than *Cs*; here, its base is at about 5000 metres, and it may well continue to lower and thicken, becoming *Ns* as shown in Plate I. Below the *As* are ragged *Cu* clouds, bases at about 800 metres and tops at about 1500 metres.

PLATE VII

1 ↓   2 ↓   3 ↓

2→

1→

←3

D. E. Pedgley

Cumulus and cumulonimbus

A dense mass of cloud is forming in a moist unstable atmosphere. The typical dark flat base is very well marked; here it is at about 1000 metres. At 1 the top of a large growing tower of *Cu con* has already built up to about 8000 metres, whereas at 2 another tower has reached about 10 000 metres and, although its outline still shows a 'cauliflower' appearance, the side has far less detail than has 1—it is changing to *Cb*, species *cal*. This change starts as glaciation sets in. At 3, a third tower has stopped growing and has developed a fibrous top as a result of more advanced glaciation—*Cb cap*. In the foreground are some patches of *Cu fra*.

PLATE VIII

D. E. Pedgley

Cumulus and stratocumulus

Convection near the ground is producing *Cu* with dark flat bases at about 1200 metres, but a stable layer near 2000 metres is preventing their upward growth so that the cloud tops spread sideways as patches of *Sc cugen.* The base of the *Sc* shows *mamma.*

PLATE IX

D. E. Pedgley

Cumulus and cumulonimbus

The top of a vast *Cb* cloud, extending up to about 10 000 metres, is streaming away to the right as a result of strong winds aloft. The oldest part of the cloud is at the top right and is quite fibrous—*Cb cap*. To the left the top has an edge which is more clear cut, but the whole of the cloud facing the observer is typically smooth and lacking in detail, indicating that convection has slackened and that growth of precipitation within the cloud is well advanced. In the foreground are *Cu* clouds, species *med* and *fra*, and also some *Ac cugen*.

PLATE X

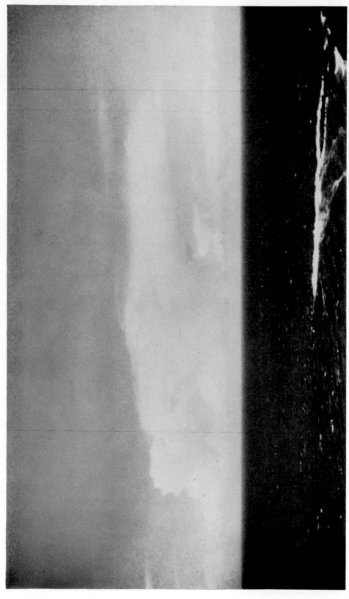

D. E. Pedgley

Cumulonimbus

The massive *Cb* cloud (side view) is about 50 kilometres away. Vertical wind shear from left to right is distorting the cloud into an anvil; its flat top indicates where the tropopause is preventing further upward growth. New growth occurs at the upshear (left-hand) end.

PLATE XI

Cumulonimbus

The underside of a retreating *Cb* anvil shows *mamma*.

PLATE XII

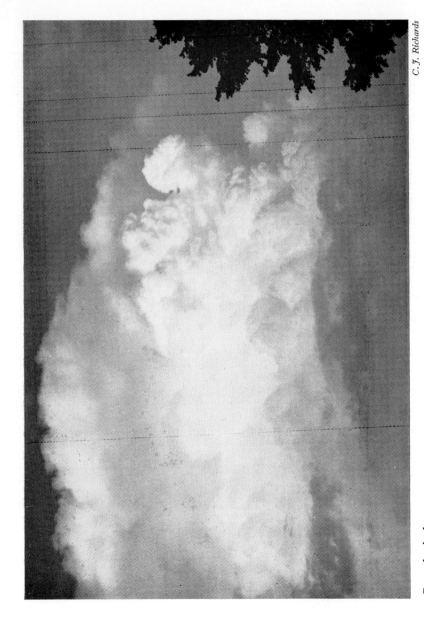

C. J. Richards

Cumulonimbus

Heat thunderstorm after a fine, hot day. Vigorous convection is evident within the cloud mass, with a recently formed anvil canopy spreading out from the cloud top.

PLATE XIII

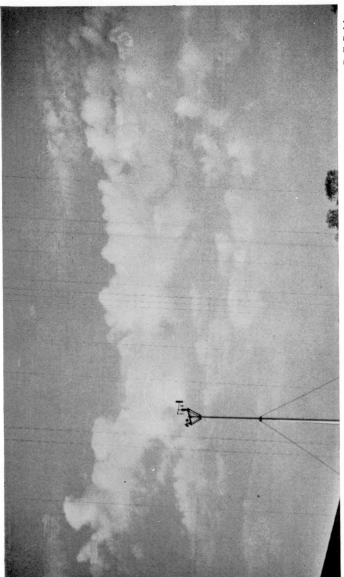

D. E. Pedgley

Altocumulus

    This *Ac* differs in an important way from the *Ac* shown in Plates II and V, and from the *Sc* in Plate IV. Here it is distinctly cumuliform with bulging tops but rather ragged bases. Some of the patches are isolated (species *flo*) whereas some tend to be in rows with a common base (species *cas*). Both types indicate instability at medium levels, here at about 3000 metres.

PLATE XIV

Cirrus

Two types are present. In the centre is species *cas*, with a flat base and cumuliform tops and with *virga* beneath. Elsewhere the *Ci* is dense but in irregular patches—species *spi*. The cloud is at about 7000 metres, where the atmosphere is unstable as shown by the *Ci cas*.

PLATE XVI

D. E. Pedgley

Altocumulus

The cumuliform tufts of *Ac* are similar to those in Plate XIV but higher, about 5000 metres. At this level the temperature is low enough to allow some ice particles to form in the otherwise supercooled clouds. These particles grow and fall out as streamers of cirriform cloud known as *virga*, but they sublime after falling about 1000 metres. In the foreground are some *Cu fra*.

PLATE XV

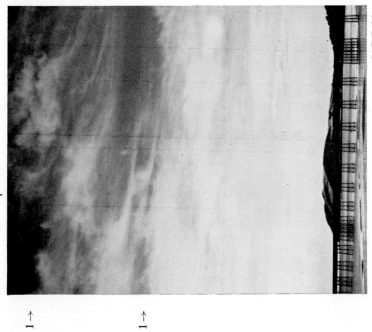

1 →

1 →

D. E. Pedgley

Cirrus and cirrostratus

In the upper part of the picture are tufts of *Ci* with streamers of ice particles trailing beneath them (species *unc*), best shown at 1. Towards the horizon the cloud is continuous, forming *Cs*. Some *Ac len* and *Cu med* is also present. *Ci* and *Cs* together are typical of the leading edge of high cloud associated with an advancing warm front.

PLATE XVII

D. E. Pedgley

Cirrus

Thin, fine, fibrous *Ci* (species *fib*) is spreading across the sky at high levels, probably above 10 000 metres. Below are tufts of *Cu fra*.

PLATE XVIII

maintain saturation, condenses on to the special nuclei in the air as was explained in Section 3.3.2. This cooling may be brought about in three* ways:

(a) *Adiabatic expansion*, when air rises in the atmosphere to levels with a lower pressure. The rate of cooling is, of course, at the dry adiabatic lapse rate until it becomes saturated, and thereafter condensation occurs and cooling is at the saturated adiabatic lapse rate; the cloud base marks the change-over.

(b) *Contact with a cold object*, especially the ground, when its temperature is below the dew-point of the air.

(c) *Mixing* of two nearly saturated masses of air of markedly different temperatures.

Of these three methods the first is by far the most important; the second can produce cloud only near the ground, up to say 500 metres; the third method is of small value and probably never alone causes natural clouds to form (but see Section 5.6.1).

5.2.2 *Dispersal of clouds*

A cloud may disperse in two ways:

(a) *Evaporation*, either resulting from a rise of temperature causing the relative humidity to fall below 100 per cent (which can be brought about simply by reversing two of the processes of formation, that is, by contact heating and by adiabatic compression accompanying descent in the atmosphere) or by mixing with its dry surroundings, especially for cumuliform clouds since they are relatively isolated.

(c) *Fall-out* as precipitation. A large fraction of the water in those clouds which give precipitation is removed in this way.

Either of these methods can be important on particular occasions as we shall see in the next two sections. Since adiabatic expansion and compression are so important in the mechanisms of cloud growth and decay it will be useful here to restate the various types of vertical motion which bring them about—they were mentioned in Section 2.6—turbulence, orographic ascent, convection, convergence and frontal ascent. It should be noted that whereas adiabatic processes can occur at almost any level in the atmosphere, cooling and warming by contact is confined to that part nearest the ground.

5.3 STRATIFORM CLOUDS

5.3.1 *Introduction*

Stratiform clouds develop in a *stable* atmosphere. They give visual evidence of widespread cooling, usually as a result of adiabatic expansion but in some circumstances through contact with a cold ground. Stratiform clouds are typified during active growth by:

(a) a wide horizontal extent compared with their vertical depth,
(b) often diffuse outlines,
(c) feeble vertical air currents.

---

* A fourth method is also known—loss of heat by long-wave radiation—but is important only after clouds have formed; see, for example, Section 5.3.3.

It is convenient for our purposes to separate stratiform clouds into four groups:

(a) Stratus.
(b) Thin stratiform clouds of species *stratiformis* (*str*).
(c) Thin stratiform clouds of species *lenticularis* (*len*).
(d) Thick stratiform clouds and multi-layered clouds.

5.3.2 *Stratus*

Stratus is the cloud which has the lowest base of all, and indeed it is often associated with fog; only seldom has it a base above 500 metres. This low base, combined with the poor visibility often found with it, makes it of special importance to aviation: it is one of the major weather factors which determine the usefulness of an airfield. Even this apart, the fact that it gives dull weather, often persistent and widespread, makes it of interest to the public too.

Probably *St* forms most commonly by advection in much the same way as advection fog forms (see Section 4.3), that is, when moist air blows across a cold surface whose temperature is below the dew-point of the air. This is illustrated in Figure 22. Two factors decide whether *St* or fog will form, or whether no

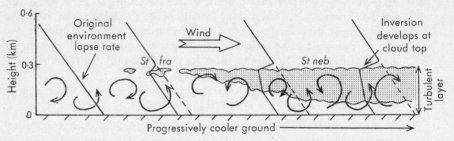

FIGURE 22. *Formation of advection stratus cloud when moist air flows over cold ground*
Changes in lapse rate shown.

condensation at all occurs: these factors are wind strength and the temperature difference between air and ground. A wind above about 5 to 10 knots favours *St*, but if it is too strong the sky may remain clear because the cooling is spread through too deep a layer of the atmosphere thus giving a temperature fall insufficient to cause condensation. With a large temperature difference between air and ground the effect is the reverse, so that even with strong winds *St* forms. When condensation does occur it is at the top of the mixed layer. A thawing snow surface is very effective in producing *St* when moist maritime air flows across it. Two favourable regions in Britain for advection *St* are:

(a) The west and south coasts when warm, moist south-west winds blow. As the air approaches Britain it crosses a progressively cooler sea surface and becomes filled with extensive shallow *St* which then drifts on to all exposed coasts and often well inland too.
(b) The east coast when moist east winds blow. The air is cooled by the cold North Sea and is most likely to produce advection *St* in early summer when the temperature contrast is greatest; this is because the air flows

from a warm continent, but the sea has only warmed a little since the previous winter. In Orkney such persistent low *St* is known as 'haar', and the use of this name has now been extended to cover all such *St* on our east coasts. It is also known as 'North Sea stratus'.

Advection *St* usually disperses through heating which causes its evaporation. Day-time sunshine usually produces the heating, but occasionally the cloudy air may flow over a warmer surface. An increase in wind speed can disperse the cloud but the process is complex. Since advection *St* is formed by cooling from below, the atmosphere near the ground is stable and there is often an inversion with its base at, say, 200 to 1000 metres. The cloud's top is often well defined and near the inversion base, so that high ground may extend above it into clear air. The cloud's base usually shows little or no detailed structure—species *nebulosus* (*St neb*); when broken, that is, when just forming or dispersing, it is known as the species *fractus* (*St fra*).

Other ways by which *St* may form are:

(a) The passage of warm, moist air over ground cooled at night by loss of long-wave radiation, especially after previous rain which has wetted the ground. This method is usually found only as an aid to the formation of advection *St*.

(b) Orographically, especially over our hilly western and northern districts which may become enveloped in this cloud (see Plate I).

(c) By turbulence, when precipitation is falling from a cloud layer at a higher level. As the precipitation falls, some may evaporate in the air below the cloud which thus becomes moistened and cooled. Air near the ground which becomes nearly saturated in this way may produce fragments of cloud in the irregular eddies accompanying turbulence, and as the process continues the cloud becomes more extensive, perhaps eventually giving a complete cover which merges with the other layer above. This type of *St* is known as *pannus* (*pan*); it is quite commonly found with *Ns*, *As*, *Cu* and *Cb*.

(d) As 'lifted fog'. In Section 4.2.1 we saw that when fog disperses it may do so through a temporary stage of low *St*, and on some days this may persist.

5.3.3 *Thin stratiform clouds of species stratiformis (str)*

Each of the three genera *Cc*, *Ac* and *Sc* exists in large numbers of species, the most common being *stratiformis* (*str*) indicating an extensive horizontal sheet. When these clouds are present, their detailed structures and pronounced variations in both colour and shading give a very pleasant appearance to the sky. They are almost always formed in air rising slowly as a result of convergence, especially in the vicinity of fronts, and they usually disperse by a reversal of this process, that is, by adiabatic warming through slow descent.

There are several varieties of the species *stratiformis*, some of which can coexist:

(a) Variety *perlucidus* (*pe*), in which the sheet is broken into a large number of cloudlets with at least some distinct spaces between them thus giving a characteristic 'crocodile-skin' or 'crazy-paving' pattern illustrated in Plate IV. After the cloud sheet first forms, long-wave radiation from its top, outwards into space, results in cooling (even by day) so that the lapse rate within the cloud layer increases until it finally exceeds the saturated adiabatic lapse rate. Very shallow convection then sets in, producing the distinctive *Bénard cells*, the air in each of which slowly overturns resulting in a thickening of the cloud in the central upcurrents but dispersal on the edges where the air descends. The convection is prevented from extending upwards to levels above the cloud top by a stable layer which develops there, also a result of the cloud's cooling (see Figure 23). Sometimes the *pe* variety may be seen to form directly

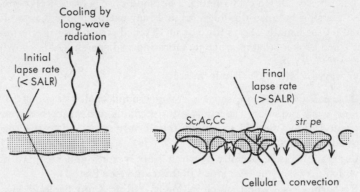

FIGURE 23. *Formation of variety perlucidus in a thin stratiform cloud*

from clear air, probably because a cellular pattern in the distribution of humidity still persists for a while after the dispersal of a similar cloud from that air. The alternating periods of up and down motion necessary to bring about such a disappearance and reformation of the cloud are not uncommon, especially near weak depressions and fronts. Careful observation shows that the size of the cloudlets increases with the depth of the overturning layer. At high levels the cloud is thin and so its cells are small. At medium levels the cells in *Ac str pe* are larger, but they can be quite small when the cloud first forms so that the newest edges of a steadily growing sheet may have a fine structure resembling *Cc* whilst the older parts are more coarsely divided. With *Sc str pe* the cells are larger still.

(b) Variety *translucidus* (*tr*), similar to *pe* but the sun or moon can be seen through the cloud (not just between the elements) because it is so shallow.

(c) Variety *lacunosus* (*la*), a rare form having a 'net-like' or 'honeycomb' structure. The clear spaces present in the cloud sheet appear to be the result of cellular convection of cloud-free air into the body of the cloud, either from above or below.

(d) Variety *undulatus* (*un*), in which the sheet is divided into approximately equally spaced and parallel bands, the spacing being about 500 to 1000 metres, and their axes being often approximately across their direction of motion. Examples are shown in Plates II and III. These bands, or *billow clouds*, form in the crests of a series of *shear waves* which sometimes appear at the horizontal boundary surface between two superimposed layers of air moving at different velocities, the clouds themselves moving at an intermediate velocity (see Figure 24). A favourable place is near a jet stream.

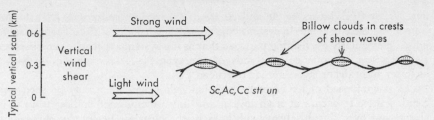

FIGURE 24. *Formation of variety undulatus in a thin stratiform cloud*

5.3.4 *Thin stratiform clouds of species lenticularis (len)*

Another species in which each of the genera *Cc*, *Ac* and *Sc* can exist is *lenticularis* (*len*), indicating a cloud of lens or almond shape. Lenticular clouds are very characteristic in appearance and are typically associated with the standing waves, or lee waves, set up over and to the lee of high ground when the wind blows across it (see Section 2.6.1). If, initially, a moist layer is present in the airstream, then after lifting it may give a cloud over, but well above, the mountain top. The cloud is lens-shaped and appears to be *stationary* at the crest of the rising air, but in fact the wind blows through the cloud which is continuously forming at the upwind end and dispersing downwind. Several superimposed lenticular clouds may be present. They are best formed in a stable atmosphere where the air has to be forced over the hill. Some of the crests in the series of lee waves may also be shown up by lenticular clouds (Figure 25) and such clouds are locally very well known near high ground. For example, the 'helm bar' forms over the Eden valley, lying to the west of Cross Fell in Cumbria, when an east wind blows. Lenticular clouds near the Isle of Man are shown in Plate V.

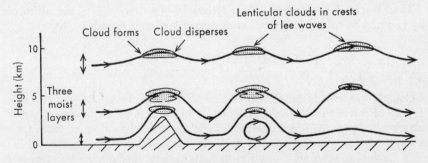

FIGURE 25. *Formation of lenticular clouds in a series of lee waves*

If conditions are favourable, lenticular clouds may also show the same structural varieties which were mentioned in Section 5.3.3. Some lenticular clouds may be seen drifting with the wind but at a lower speed; they are usually short lived whereas the others may persist apparently almost unchanged for an hour or more.

5.3.5 *Thick stratiform clouds and multi-layered clouds*

Clouds associated with our widespread and persistent rain areas are normally of this type and are often found within developing depressions, and especially near fronts, where the necessary lifting for condensation is found. These clouds are either the very extensive forms with little or no detailed structure, *Cs*, *As* and *Ns*, or thick layers of *Ac* and *Sc str* of the variety *opacus* (*Ac* (*Sc*) *str op*). The cloud base of a single system considered as a whole is often highest at its edges, particularly the leading parts, so that as the system approaches an observer he often sees a sequence of clouds of the type *Cs–As–Ns*, or *Ac–AcAs–Ns*, as lower and thicker layers spread overhead.

Cs is composed of ice crystals in which halo phenomena may be seen (see Section 8.1). It occurs at high levels, and any precipitation falling from it is slight, and over Britain always sublimes in the clear air below the cloud long before reaching the ground. As shown in Plate VI it forms a tenuous film over the sky.

As may be composed of either water droplets or ice particles. The latter are often large, in the form of snowflakes, so that even with a deep cloud, say 1000 metres or more, the sun's disc is still visible—variety *translucidus* (*As tr*)—but it is fuzzy as though it were seen through ground glass (see Plate VII). The cloud may become so thick that it obscures the sun completely—variety *opacus* (*As op*). Haloes are not seen because the crystals are not of the right type. Slight rain or snow may reach the ground from *As*; otherwise, any precipitation evaporates before reaching the ground.

Ns is just an advanced stage of *As*. It has a great vertical depth, often several kilometres and sufficient to obscure the sun (Plate I), and its base extends down to low levels. Continuous precipitation usually falls from it.

Instead of one thick layer of cloud, several thinner layers may be present at the same time. Such combinations are typical of the outer parts of the vast cloud systems associated with depressions, and they will be discussed further in Chapters 9 and 10. A common combination is: *Cs* and *As* sheets above with broken layers of *Ac* below. Any precipitation from such clouds is intermittent and slight.

Thick layers of *Ac str op* and *Sc str op* consist of water droplets, and although they give persistently cloudy weather any precipitation is usually slight.

5.4 CUMULIFORM CLOUDS

5.4.1 *Introduction*

Cumuliform clouds develop in an *unstable* atmosphere. They give visual evidence of convection in the atmosphere, and the cooling responsible for condensation is always the result of adiabatic expansion in the rising currents. Once the air starts rising it continues to do so spontaneously: all that is needed is a 'trigger action' to set if off. Cumuliform clouds are typified during active growth by:

(a) a heaped appearance with swelling, domed tops corresponding to the rising air currents,

(b) clear-cut outlines and more or less horizontal bases,

(c) vigorous vertical air currents.

By far the most important cumuliform clouds are cumulus and cumulonimbus, but they can also occur at medium and high levels (Section 5.4.4).

5.4.2 *Cumulus*

Cumulus clouds are among the commonest clouds observed; they are well known as the 'cotton wool' clouds of a summer day. A great interest has been taken in them during recent years largely as a result of their importance as sources of aircraft icing, of 'lift' for gliders, and for their possible use in the artificial production of rain. Our knowledge of *Cu* clouds has made dramatic advances but there is still much to be learnt.

Probably the most common way by which they form is through the irregular heating of the lower atmosphere by contact with a warm underlying ground, itself warmed by insolation. The heating of the air is irregular because the ground temperature is not uniform—it varies markedly from place to place for reasons that were discussed in Section 1.5.2, so that bare sand and concrete, for example, are much hotter by day than neighbouring areas of grass or forest. The warmest spots, known as *thermal sources*, become covered by air which is warmer than its surroundings and which, because it is less dense, tends to rise in the form of 'bubbles', known to glider pilots as 'thermals'. A series of bubbles starting in this way makes up a convection current in the atmosphere. The bubble theory of convection has done much to explain the properties of *Cu* clouds, and foremost amongst its investigators have been Scorer and Ludlam in Britain and Malkus in the United States of America.

A bubble of warm, buoyant air (Figure 26) consists of:

(a) A *cap*, representing the uppermost, leading portion of the warmest air, and having the greatest temperature contrast with its environment.

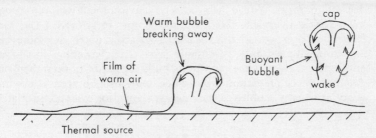

FIGURE 26. *Formation of convective bubbles by heating the lower atmosphere*

(b) A *wake*, consisting of less buoyant air resulting from the progressive mixing of the cap and sides of the bubble with its environment. This mixing process, which causes a progressive diminution of the cap's size together with an accompanying increase in the size of the wake, is known as erosion.

After rising a distance equal to about once or twice its own diameter, the bubble is spent; new and larger bubbles form by the combination of wakes of smaller bubbles, so that in general the further one goes above the ground the larger are the bubbles, but also the smaller is the temperature difference between a bubble and its environment. Whilst a rising bubble remains unsaturated, the environment lapse rate must exceed the dry adiabatic lapse rate if the atmosphere is to be unstable (see Section 1.4.2); such a condition is common near the ground during the middle part of the day. Eventually, some of the rising bubbles cool sufficiently to reach saturation, after which further ascent produces clouds. They are small, ragged and show the cumuliform characteristics only poorly developed, so they are known as the species *fractus* (*Cu fra*). Further development arises from the joining of smaller cloudy bubbles, just as the cloud-free bubbles join below the cloud base (Figure 27(a)). However, there are several factors controlling the buoyancy of these bubbles, and hence also the sizes and shapes of clouds formed in them:

(a) *Stability* of the environment above the condensation level. Since the bubbles are saturated the environment lapse rate need only exceed the saturated adiabatic lapse rate (and not necessarily the dry adiabatic lapse rate) to make the atmosphere unstable. Such a situation of conditional instability (Section 1.4.2) is normally present above the surface layers on a day when *Cu* forms. Thus, once a small fragment of *Cu fra* appears we would expect it to continue rising until it reaches some higher level where its buoyancy is destroyed. This would occur, for example, when the cloud top penetrates into a stable layer aloft.

(b) *Mixing* with its surroundings (a process known as *entrainment*). Whilst a rising bubble remains unsaturated, mixing produces a wake which is still somewhat buoyant, but mixing of a cloudy bubble with clear, unsaturated surroundings gives it a cold, dense 'skin', because evaporation of the cloud droplets absorbs latent heat so cooling the boundaries of the cloud until they become denser than either the cloud's centre or its surroundings. The cloud boundaries sink, giving a 'fountain effect' which can be seen quite clearly on the surfaces of *Cu* building up into a dry environment. We see then that a dry environment with rapid entrainment acts as a very effective break to the upward growth of *Cu*; without it we would expect all the *Cu* clouds to grow up to the same height, and to be stopped only by a stable layer aloft. Variations in the rate of entrainment allow some clouds to build upwards farther than others. Another effect of entrainment which may also be seen is the ragged edges of cloudy bubbles as they decay, shown for example in Plate VII

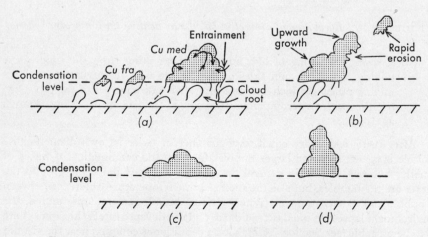

FIGURE 27. *Some factors controlling the growth of Cu clouds*

(a) Formation of *Cu* when convective bubbles rise above the condensation level.
(b) Effect of vertical wind shear.
(c) Effect of slight instability or dry environment.
(d) Effect of great instability or moist environment.

(c) *Vertical wind shear*. If the wind velocity changes with height the clouds become distorted from the vertical as they grow upwards. For example, if the wind speed increases with height whilst its direction does not change, then a *Cu* cloud will lean over downwind (Figure 27(*b*)). This leaning also favours entrainment.

(d) *Friction* between the rising bubble and its environment. This force effectively balances the bubble's buoyancy so that it rises at a constant speed.

(e) *Accumulation of water and ice*. The conversion of water vapour into the much denser water and ice as a result of condensation within the cloud, leads to an increase in the density of the cloud as a whole, and since the effect increases as the cloud builds higher, it can alter the buoyancy considerably.

Summarizing, we may now picture a *Cu* cloud which forms by surface heating as consisting of:

(a) a stream of unsaturated bubbles rising from the ground to cloud base (known as the 'cloud root');

(b) several larger bubbles within the cloud, buoyant and tending to surge upwards, but controlled by entrainment, wind shear, friction and accumulation of water;

(c) a descending 'skin', the result of cooling by mixing at the edges.

As a new bubble emerges from the cloud top, it erodes and leaves a wake which is moister than the environment. Such a wake is favourable to a greater upward penetration of the next bubble to emerge, and in this way the cloud slowly builds upwards in irregular steps. Its life ends when the supply of fresh bubbles entering its base is cut off; rapid mixing follows and the cloud soon disappears. The lifetime of a *Cu* cloud is very variable but is roughly between 5 and 30 minutes; the larger the cloud the longer its life. The combined effects of stability and entrainment determine the relative dimensions of the cloud—slight instability or a dry environment both favour broad and squat clouds (Figure 27(*c*)), whilst great instability or a moist environment favour tall and narrow clouds (Figure 27(*d*)—see also Plate VIII). Incidentally, very similar effects may be seen in steam plumes from chimneys of locomotives and power-stations.

When a *Cu* cloud succeeds in building up to a stable layer a further interesting effect may be observed. The buoyant bubbles are incapable of ascending through the stable layer so they accumulate just below it and spread sideways to form an area of stratiform cloud, known as stratocumulus (or altocumulus) *cumulogenitus* (*Sc* (*Ac*) *cugen*) as shown in Figure 28 and Plate IX. Such clouds are lumpy and usually thicker and darker than *Sc* (*Ac*) *str*, and since they are relatively stagnant compared with the parent *Cu* (upward motion in them is small) they decay by mixing more slowly; they often persist in patches for an hour or more after the parent *Cu* has dispersed. When vigorous *Cu* spreads below such a stable layer, the patches of *cugen* cloud may combine to give a continuous and persistent layer which reduces insolation, so hindering the formation of new *Cu*. On some days which start clear, *Cu* develops during the morning but by the afternoon it is persistently cloudy because *Sc* or *Ac cugen* has accumulated. The mixing of this layer cloud with its surroundings takes place mostly on its undersurface, and the resulting cold 'skin' descends into the clear air below in the form of 'drops'

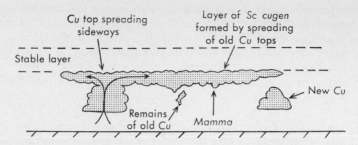

FIGURE 28. *Formation of Sc cugen when Cu tops spread below a stable layer aloft*

which show up on the underside of the cloud as a festoon structure known as *mamma*. This feature can also be seen occasionally on other forms of *Ac* and *Sc*, especially *Ac* (*Sc*) *str op*. It is an example of downward convection in the atmosphere and, since this is accompanied by adiabatic warming, the *mamma* soon evaporate; most *mamma* do not descend more than a few decametres.

If the stable layer which limits convection lies just above the condensation level then the *Cu* clouds are very shallow and are known as species *humilis* (*Cu hum*) (Figure 29 and Plate VI); with increasing vertical extent we have species *mediocris* (*Cu med*), perhaps about a kilometre deep as in the foreground of Plate X, and finally, large, towering clouds of species *congestus* (*Cu con*) which may be 2 to 3 kilometres deep or more, as in Plate VIII.

Occasionally, a smooth 'cap' of cloud may be seen on or just above the top of a vigorous *Cu*. This is caused by a localized lifting of the environment above the swelling *Cu* top. If it is moist enough, these surroundings develop their own cloud known as *pileus* (*Cu pil*) (Figure 30); it is short-lived, for the *Cu* either pushes up through it or it subsides back to lower levels only to evaporate as a result of adiabatic warming.

When the unsaturated bubbles can hardly reach the condensation level, their presence in a smoky atmosphere may be shown by the beginnings of condensation, giving misty patches in the same way that mist forms before a smoke fog even though the air is not saturated. These 'fumulus' clouds may grow into normal *Cu*.

The height at which condensation takes place in a rising bubble (that is, the cloud base) depends upon its humidity when leaving the ground. If the humidity is low, much cooling is needed to reach saturation, so the base is high;

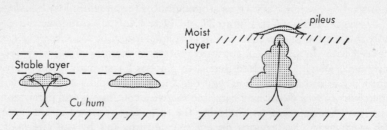

FIGURE 29. *Shallow Cu hum forming below a stable layer*

FIGURE 30. *Pileus forming in a moist layer of the environment pushed up by a vigorous Cu*

conversely, a high humidity gives a low base. This explains the following easily observed facts concerning *Cu* bases:

(a) During daylight hours the base rises, reaching a maximum in mid afternoon corresponding to the diurnal fall of relative humidity near the ground. Thus, when first forming they may be at 500 metres, progressively rising to, say, 1200 metres or even 2000 metres.

(b) Bases are generally lower in winter than in summer.

(c) Bases are lower on windward coasts than inland.

We are now in a position to explain the well-known inland *diurnal variation* of *Cu* development. After a clear night there is a stable layer in the lowest part of the atmosphere (see Section 1.5.2) which is destroyed during the morning by insolation heating the ground, so that the atmosphere becomes unstable over a gradually increasing depth. *Cu fra* appears in due course, tending to grow upwards but controlled by the factors already considered. By late afternoon or early evening the supply of fresh bubbles ceases because the lowest layers have become stable again, and the existing clouds are eroded away so that the sky becomes clear or only a few patches of the more persistent *Sc* (*Ac*) *cugen* remain. It should be noted that the clouds present, say, at mid morning are not the same as those seen in mid afternoon; this is not unexpected when we remember that the life-cycle of an individual cloud is quite short.

Much has still to be learnt about the way *Cu* clouds are distributed over the landscape. They form best over good thermal sources, from where they drift downwind to give no obvious pattern in the sky. On some days, however, rows of clouds lying along the wind can be seen clearly; they are known as 'streets', or variety *radiatus* (*Cu ra*). *Cu* forms better over land than over lakes or the sea, where it shows little diurnal variation. In hilly country, sun-facing slopes act as good thermal sources so that over them *Cu* forms earlier and disperses later than over flat country or shaded slopes.

On some days the normal diurnal cycle of *Cu* development is modified or not observed. Some causes of this are:

(a) The environment is too dry, so that entrainment maintains too low a humidity in the rising bubbles.

(b) A stable layer may be present whose base lies below the condensation level; this is especially true of anticyclones in summer.

(c) Cloud is present at a higher level, so reducing insolation.

Cu may also occur at unexpected times, for example at night. This is nearly always the result of convection being released by a 'trigger action' other than the normal one of irregular heating of the atmosphere by the ground. Thus convergence, frontal ascent and orographic ascent may produce *Cu* at any time as long as the atmosphere is unstable. Such clouds have no 'roots', their bases are smoother in appearance and more uniform, and they show a distinct tendency to form in groups or lines rather than being scattered, apparently at random, over the sky.

5.4.3 *Cumulonimbus*

Cumulonimbus clouds normally develop from large cumulus. Their appearance differs from *Cu* in that the upper parts have lost their typical sharp 'cauliflower' outlines: the tops become diffuse and striated. These changes are a visual indication of the development of extensive precipitation within the cloud, often accompanied by a slowing down or even reversal of the upward currents. Precipitation in the upper part of the cloud is usually solid (see Section 6.4.3) because its temperature is well below 0 °C. *Cu* may sometimes be seen to change to *Cb* when the temperature of its top falls below about −10 °C but usually temperatures below −20 °C are necessary, and since these values are found only at some considerable height above the ground—usually 4 to 6 kilometres—*Cb* clouds have great vertical development. When the whole depth of the troposphere is unstable, *Cb* tops may tower up to the tropopause; tops above 12 kilometres have been observed in Britain, but in equatorial regions they can sometimes exceed 18 kilometres. Such clouds are usually accompanied by heavy precipitation and sometimes thunder and lightning too. In cold weather much shallower *Cb* can form, sometimes less than 3 kilometres deep, because temperatures of −10 °C and below are then found much lower in the atmosphere.

The transition from towering *Cu* into *Cb* needs temperatures of −10 °C and below because it is only with such low values that natural freezing nuclei become effective (see Section 3.4.1). Once it starts, the transition is rapid; Figure 31 shows a typical sequence of changes. In about 10 to 20 minutes, much of the cloud above the 0 °C level is transformed from supercooled water droplets into ice particles—a process known as glaciation. After glaciation the cloud top may degenerate into dense masses of fallstreaks, giving a cirriform cloud known as *cirrus spissatus cumulonimbogenitus* (*Ci spi cbgen*), that is, dense *Ci* formed

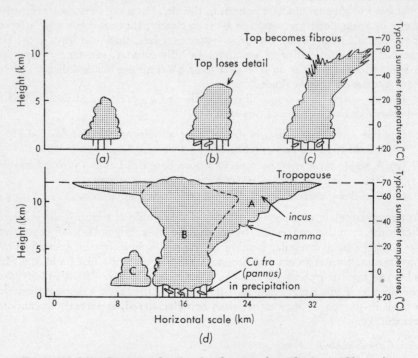

FIGURE 31. *Sequence showing development from Cu con to Cb cap inc*

(a) *Cu con.*   (b) *Cb cal.*   (c) *Cb cap.*   (d) Composite *Cb cap inc.*

A: Remains of old cell,   B: New cell,   C: First stage of new cell.

from *Cb*. However, further large bubbles often surge upwards from the interior of the cloud so extending the top to higher levels. A *Cb* cloud which has developed a fibrous cirriform top is known as species *capillatus* ('hairy', *Cb cap*)—see Figure 31(*c*); the transition period, when the *Cu* is changing to *Cb*, is marked by the cloud top becoming smooth as it loses its structural detail, and the *Cb* species is then *calvus* ('bald', *Cb cal*)—see Figure 31(*b*). Plate VIII shows *Cb cal* and Plate X *Cb cap*.

Where the top of a developing *Cb* meets a stable layer aloft, it spreads sideways just beneath it, but the resulting cloud is different from that produced by the similar spreading of a *Cu*. It is smooth with fibrous edges and resembles *As* or *Ns*; indeed, the accumulation of such cloud from several large *Cb* may be called *As* or *Ns cbgen*—it is a layer cloud formed by the spreading of *Cb* tops. Its appearance is different from *Ac* or *Sc cugen* because the latter is composed of water droplets and seldom gives precipitation, whereas the former is composed largely of ice particles, especially snowflakes, and often gives precipitation which causes the cloud's base to be diffuse and lacking in detail. When a spreading top is still part of its parent cloud, it resembles an anvil (variety *incus* (*Cb inc*)— Figure 31(*d*)) especially when drawn out by vertical wind shear, well shown in Plate XI. Such wind shear tends to maintain the growth of a *Cb* cloud, so that if, for example, the wind speed increases with height, the top streams away downwind as a great plume of *As cbgen* and *Ci spi cbgen*, often for many tens of kilometres, whilst new convection continues at the upwind end (Figure 32). This can be seen in Plates X and XI. The underside of an anvil may develop *mamma* in the same way that *Sc* (*Ac*) *cugen* does; in fact, the best displays of *mamma* are associated with *Cb* clouds. An example is seen in Plate XII.

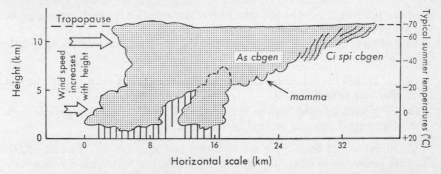

FIGURE 32. *Growth of Cb when vertical wind shear is present (wind speed increases with height)*

After some hours, an apparently simple *Cb* cloud may really consist of the remnants of several clouds which have formed earlier, together with more recent developments, the whole forming a complex cloud system which can cover a wide area (Figure 31(*d*)). When convection ceases *Cb* clouds decay, leaving large irregular masses of cloud at many levels (Figure 33), especially *Ci spi cbgen* which is the most persistent, perhaps remaining for 12 or even 24 hours before dispersing by fallout and sublimation. In a very unstable troposphere the accumulation of these persistent tops may prevent the growth of new *Cb* by reducing sunshine reaching the ground.

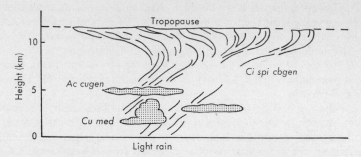

FIGURE 33. *Clouds remaining after decay of Cb cap inc*

The great vertical depth through which the convection takes place inside a *Cb* cloud results in powerful updraughts and downdraughts. Vertical speeds in excess of 30 knots have been recorded, presenting a formidable source of danger to aircraft. Further discussions about *Cb* clouds will be found in Section 6.4.3 and in Chapter 7.

5.4.4 *Cumuliform clouds at medium and high levels*

So far, all the cumuliform clouds that have been considered have had their bases low down in the troposphere. Indeed, by far the majority are of this type, but occasionally one can see clouds at medium and high levels which have definite cumuliform characteristics. They are not produced by irregular heating of the atmosphere by the ground, as can be easily understood when it is realized that their bases are almost always considerably higher than the condensation levels of surface air. The 'trigger action' which sets off convection is usually convergence near a depression, but sometimes it is orographic ascent.

Cumuliform clouds at medium levels take the form of two characteristic species of *Ac*:

(a) *Ac castellanus* (*Ac cas*) resembling a row of turrets with a well-defined flat base (Figure 34(a)).

(b) *Ac floccus* (*Ac flo*) broken into irregular tufts having cumuliform tops but ragged, eroding bases (Figure 34(b)).

Combinations of both types are common and may be seen in Plate XIV. They are indicators of instability at medium levels, but when first forming they may show no signs of it, rather do they resemble patches of *Ac len*. However,

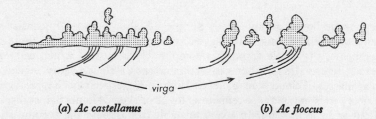

(a) *Ac castellanus*   (b) *Ac floccus*

FIGURE 34. *Two species of Ac which form in an unstable atmosphere*
Both show precipitation trails (*virga*).

within minutes cumuliform tops sprout upwards. Sometimes the cloud becomes extensive, deep and almost continuous so that it is difficult to see the towering tops, but it is still easily identifiable by a characteristic mottled pattern in the detail which appears on its underside and which shows up the denser parts by areas of darker shading. These clouds may be found at several levels simultaneously and also coexisting with stratiform clouds, both *Ac* and *As*; the whole sky then takes on a chaotic appearance.

We have seen that instability is essential to the growth of *Cb* clouds, and since *Ac flo* and *Ac cas* indicate instability in the middle troposphere, their appearance is often an indicator of probable development of *Cb* and thunderstorms in the vicinity of the observer, say, within 100 miles. The more extensive and denser forms can themselves give outbreaks of rain and thunderstorms (see Sections 6.4.4 and 7.3.1). Trails of precipitation resembling cirriform clouds may sometimes be seen descending from cloud base and evaporating before reaching the ground. These trails are known as *virga* and can be seen in Plate XV; when well developed they transform the appearance of a massive cell of *Ac flo* into that of a miniature *Cb*. *Ac flo* occurring at low temperatures (usually below −20 °C) may sometimes be seen to change completely into fallstreaks of ice particles, producing a species of *Ci* known as *altocumulogenitus* (*Ci acgen*), taking some 10 to 20 minutes to do so.

Cumuliform clouds at high levels may sometimes be seen but they are rare. The very low temperatures at these levels usually ensure that any water-droplet cloud which might form changes quickly to ice particles. The *Ci* species *Ci floccus* (*Ci flo*) and *Ci castellanus* (*Ci cas*) resemble their *Ac* counterparts but show finer detail and are almost always accompanied by other *Ci* species. Plate XVI illustrates this.

5.5 CIRRIFORM CLOUDS

5.5.1 *Formation and properties*

Cirriform clouds are fine streamers of precipitation in the upper troposphere. When freshly formed they are clear-cut and often have well defined 'heads' below which trails of ice particles descend; old clouds are more diffuse and irregular. They are composed of ice particles, mostly crystals, and are formed either by transformation of a pre-existing water-droplet cloud (for example, *Cc* or *Ac*, as was seen in Section 5.4.4) or more commonly by apparent direct deposition of water vapour from clear air. Even in the latter event, however, it is possible that the first condensation produces a few minute water droplets which, being greatly supercooled, soon freeze and subsequently act as centres for further deposition of water vapour. The low temperatures high up in the troposphere keep the saturated vapour pressure very low, and this, together with only a slow release of water vapour as the air in which the cloud forms ascends, ensures that the cloud is very tenuous, so tenuous that the sun or moon is nearly always visible through it.

Most cirriform clouds form in air rising slowly as a result of horizontal divergence aloft near developing depressions (see, however, the *Ci* formed from *Cb* discussed in Section 5.4.3). Their usual base height is 8 to 11 kilometres but about 10 per cent have bases above 13 kilometres. Observations from aircraft show that a given layer of *Ci* often has a very irregular base but a remarkably flat top, thus giving a variable vertical depth to the cloud. About half of the observed *Ci* tops lie within the uppermost two kilometres of the troposphere.

As the crystals in a *Ci* cloud grow they fall away from their 'generating level', eventually descending to a layer which is unsaturated with respect to ice where they sublime, and the visible fallstreaks end (Figure 35). On favourable occasions these streamers are several miles long and stretch for some distance across the sky. Their exact shapes depend on the variation of wind velocity with height. *Ci* which has distinct 'head' and 'tail' regions is known as species *uncinus (Ci unc)* but where there are only diffuse, fibrous 'tails' it is species *fibratus (Ci fib)*. These two types are shown in Plates XVII and XVIII. Denser forms are known as species *spissatus (Ci spi)*. Details on the undersurfaces of these clouds, such as *virga* and *mamma*, are best seen when illuminated by the rising or setting sun. There is still much to be learnt about *Ci* clouds.

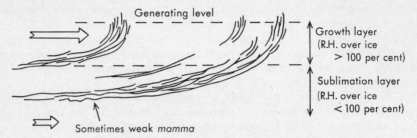

FIGURE 35. *Formation of Ci unc when vertical wind shear is present*

5.6 CONTRAILS AND DISTRAILS

5.6.1 *Contrails*

Condensation trails caused by the passage of an aircraft through the atmosphere are artificial clouds and are of two types: *adiabatic trails* and *exhaust trails*. Adiabatic trails are produced by a reduction of pressure in the immediate vicinity of the aircraft as it passes through the air. The most spectacular form is an aura which envelops much of the wing surfaces. A high speed and a high humidity are favourable to its formation. Rather more common are the short, non-persistent trails extending behind the aircraft's wing tips and formed in the series of vortices which form there and rapidly break away only to disperse by turbulent mixing with the surroundings. Similar trails can form at propeller tips.

Exhaust trails result from the condensation of water vapour contained in the engine exhaust, which is hot and moist and cools rapidly both by radiation and mixing with its surroundings. Since this cooling takes a finite time, the trail first appears some 50 to 100 metres behind the aircraft. The persistence of the trail depends upon the temperature and humidity of the atmosphere at aircraft level. Low temperatures favour ice particles which are more persistent than droplets; a moist atmosphere counteracts the dissipative effect of continued mixing. If the air is supersaturated with respect to ice, as may well occur near cirrus clouds, the particles in the trail may even grow, making the trail dense and persistent. The shape of the trail is determined by the motion of the air in which it is laid down. Several features are important:

(a) Mixing with the surroundings causes evaporation of the trail and cooling similar to the cooling found on the edges of *Cu* clouds (see Section 5.4.2). Eventually it breaks into cumuliform tufts.

(b) Downward motion produced by the aircraft. Trails formed by engines near the centre of the wing are forced downwards and outwards giving a curtain-like trail, but broken into lumps.

(c) Wing-tip vortices disperse fairly slowly and any trails entering them are protected from mixing with their surroundings, so they remain as smooth tubes of cloud. Air below and just behind the wing is swept outwards and becomes involved in the vortices, so that underslung engines are more likely to give trails entering the vortices than those which are level with the rear edge of the wing. These trails take on a wavy appearance with the separation of the wobbles approximately twice their horizontal separation. At their closest approach they are forced downwards and stretched, finally leaving trails resembling loops or rings.

(d) Vertical wind shear causes a curtain trail to be drawn out into a wide cloud. Horizontal shear may cause easily visible distortions and breaks.

5.6.2 *Dissipation trails (distrails)*

The passage of an aircraft through a natural cloud sometimes results in the formation of a lane of clear air. Favourable clouds seem to be always composed of supercooled droplets and are very thin; distrails are especially well seen in high and thin *Ac str*. They seem to be caused by the introduction of freezing nuclei into the cloud from the exhaust gases, resulting in the formation of ice particles which then fall out. If the air below the cloud is not saturated with respect to ice, the particles sublime and a clear lane is left. If the air below is saturated, however, the particles grow, giving fallstreaks which form a fibrous fringe suspended below the distrail.

5.7 STRATOSPHERIC, NACREOUS AND NOCTILUCENT CLOUDS

5.7.1 *Stratospheric clouds*

By far the majority of clouds occur in the troposphere, but occasionally reports have been received from aircraft showing the presence of clouds resembling tenuous *Cs* lying at about a kilometre above the tropopause. They are probably formed as a result of especially pronounced upward motion associated with actively developing depressions. Little is known of them, but they do not appear to be much different from ordinary *Ci* and *Cs*. Although they are not strictly stratospheric clouds, it may be pointed out that the tops of cumulonimbus associated with severe storms (see Section 7.3.2) sometimes penetrate the tropopause and extend several kilometres into the stratosphere.

5.7.2 *Nacreous clouds*

On rare occasions after sunset, clouds are visible which occur in the upper stratosphere—between about 20 and 30 kilometres. They are illuminated by the sun and often exhibit delicate pastel colouring (iridescence—see Section 8.2)—hence their alternative name of mother-of-pearl clouds. Little is known of their constitution but the fact that they appear to be stationary and occur in association with rough topography of the earth's surface suggests that they are mountain wave clouds. They have been observed over Scandinavia, Alaska and Antarctica. Because of their height they can be seen several hundred kilometres away, for example, the Scandinavian clouds can be seen from Scotland.

5.7.3 Noctilucent clouds

These clouds have been observed during the night hours of the summer months. They are very tenuous and can be seen only under suitable viewing conditions (direct illumination by sunlight against a dark sky and in the absence of lower cloud). They resemble cirrostratus in appearance but have a bluish-white to yellow colour. Measurements have shown the clouds to be at a height of 80–85 kilometres in association with the temperature minimum at the mesopause.

BIBLIOGRAPHY

BIGG, E.K. and MEADE, R.T.; 1971. High-level stratocumulus clouds. *Weather*, **26**, pp. 55–56.

CATON, P.F.G.; 1957. Occurrence of low layer-type cloud over eastern England in relation to the synoptic situation. *Met Mag*, **86**, pp. 161–169.

CORNFORD, S.G.; 1966. Stratocumulus—a review of some physical aspects. *Met Mag*, **95**, pp. 292–304.

DAVIS, N.E.; 1952. Fog and low stratus at Prestwick Airport, with notes on the diurnal variation of surface wind and temperature. *Met Mag*, **81**, pp. 231–239.

DAY, J.A. and LUDLAM, F.H.; 1972. Luke Howard and his clouds: a contribution to the early history of cloud physics. *Weather*, **27**, pp. 448–461.

FOGLE, B.; 1968. The climatology of noctilucent clouds according to observations made from North America during 1964–66. *Met Mag*, **97**, pp. 193–204.

GADSDEN, M.; 1975. The colour of noctilucent clouds. *Weather*, **30**, pp. 190–197.

HALLET, J. and LEWIS, R.E.J.; 1967. Mother-of-pearl clouds. *Weather*, **22**, pp. 56–65.

KINGTON, J.A.; 1968. A historical review of cloud study. *Weather*, **23**, pp. 349–356.

LUDLAM, F.H.; 1967. Characteristics of billow clouds and their relation to clear-air turbulence. *Q J R Met Soc*, **93**, pp. 419–435.

LUDLAM, F.H. and SCORER, R.S.; 1953. Convection in the atmosphere. *Q J R Met Soc*, **79**, pp. 317–341.

McINTOSH, D.H.; 1972. Mother-of-pearl clouds over Scotland. *Weather*, **27**, pp. 14–22.

McINTOSH, D.H. and HALLISSEY, MARY; 1975. Noctilucent clouds over western Europe and the Atlantic during 1974. *Met Mag*, **104**, pp. 180–184.

MASON, B.J.; 1969. Some outstanding problems in cloud physics—the interaction of microphysical and dynamical processes. *Q J R Met Soc*, **95**, pp. 449–485.

P[EDGLEY], D.E.; 1970. Clouds—layered clouds I and II. *Weather*, **25**, pp. 166 and 281–282.

REINKING, R.F.; 1968. Insolation reduction by contrails. *Weather*, **23**, pp. 171–173.

ROACH, W.T. and JAMES, B.F.; 1972. A climatology of the potential vertical extent of giant cumulonimbus in some selected places. *Met Mag*, **101**, pp. 161–181.

CHAPTER 6

PRECIPITATION

6.1 INTRODUCTION

6.1.1 *Classification*

In a meteorological sense, *precipitation* may be defined as particles of liquid water or ice formed within a cloud and falling towards the ground. Notice that the precipitation need not reach the ground for it can evaporate on the way down if the air through which it falls is unsaturated. Well-defined streamers of precipitation which evaporate in this way can be seen sometimes below the bases of clouds; they are known as *virga*.

We are all familiar with at least four types of precipitation—rain, drizzle, snow and hail. However, there are many more which become apparent after careful observation, so it is not surprising that an international system of classification has been devised in order that we may distinguish between them. Table IV, which is based upon the 1975 edition of the *International cloud atlas*, gives descriptions of the various forms of precipitation, together with the normal cloud types from which each form is observed to fall and reach the ground.

TABLE IV. *Descriptions of precipitation forms*

Name	Description	Normal clouds from which precipitation can fall and reach the ground
Rain	Drops with diameter > 0·5 mm but smaller drops are still called rain if they are widely scattered	Ns, As, Sc str op, Ac flo, Ac cas, Cu con, Cb
Drizzle	Fine drops with diameter < 0·5 mm and very close to one another	St, Sc str op
Freezing rain (or drizzle)	Rain (or drizzle) the drops of which freeze on impact with the ground or other objects	The same clouds as for rain (or drizzle)
Snowflakes	Loose aggregates of ice crystals, most of which are branched	Ns, As, Sc str op, Cb
Sleet	In Britain, partly melted snowflakes, or rain and snow falling together	The same clouds as for snowflakes
Snow pellets (also known as *soft hail* and *graupel*)	White, opaque grains of ice. Spherical, or sometimes conical, with diameter about 2–5 mm	Cb in cold weather
Snow grains (also known as *granular snow* and *graupel*)	Very small, white, opaque grains of ice. Flat or elongated with diameter generally < 1 mm	Sc str op or St in cold weather

87

Name	Description	Normal clouds from which precipitation can fall and reach the ground
Ice pellets	Transparent or translucent pellets of ice which are spherical or irregular, rarely conical, and which have a diameter of 5 mm or less. They are composed of frozen raindrops or largely melted and refrozen snowflakes	*Ns, As, Cb*
Small hail	Translucent pellets of ice which are spherical or irregular, rarely conical, and which have a diameter of 5 mm or less. They are composed of pellets of snow encased in a thin layer of ice	*Cb*
Hail	Small balls or pieces of ice with diameters 5–50 mm or sometimes more, falling either separately or agglomerated into irregular lumps	*Cb*
Diamond dust	Very small ice crystals in the form of needles, columns or plates, often so tiny that they appear to be suspended in the air. Invariably associated with very low temperatures	*St, Ns, Sc str op* (sometimes falls from clear air, where it is just an advanced stage of ice fog—see Section 4.4.4)

6.1.2 *Intensity and duration*

To us in Britain there is one property of precipitation which is very well known, namely, its extreme variability both of intensity and duration. The *intensity*, or rate of fall, is taken as the rate at which the precipitation would accumulate on a horizontal surface if such processes as run-off, percolation and evaporation did not occur. It is expressed in units of millimetres per hour. Rainfall intensities are usually overestimated by an inexperienced observer; in Britain most rain falls at rates of only a few millimetres per hour, or even less, and it is unusual to find a rate of fall greater than 25 millimetres per hour. In fact, the more intense the rainfall the shorter its duration is likely to be. For details of the relationship between intensity and duration see the article by E. G. Bilham in *The climate of the British Isles*, London, 1938, p. 116.

Clearly, the *duration* of precipitation must depend very closely upon the persistence of the cloud from which it falls. We may distinguish two broad types:

(a) *General precipitation*. From extensive *stratiform* clouds. It starts and stops slowly, often lasts for many hours, with or without breaks, and is generally of small intensity.

(b) *Showers*. From the relatively isolated *cumuliform* clouds. It usually starts and stops suddenly, seldom lasts more than one hour, often much less, and is of relatively large intensity.

6.1.3 *Some effects of mountains*

When an airstream flows across a mountain mass it is forced to rise and the resulting adiabatic cooling can produce extensive clouds and precipitation, but apart from this there are two other effects which we may note here. Firstly, since water is removed from the clouds over the mountains a region of relatively light rainfall, a *rain-shadow*, is found on the leeward side. Secondly, air which descends to its original level after crossing the mountains is warmer than before ascent because the latent heat which is liberated when clouds form is only partly used again to evaporate the clouds in the descending air. Only a part is used because some of the water has fallen out; the remainder of the latent heat goes to warm the air. The larger the amount of water deposited (that is, the higher the mountains or the more moist the original air) the warmer is the air after descent. This warming is known as the *föhn effect*, and to the lee of the Alps temperatures may be 2 to 5 °C warmer than would otherwise be expected. Not only is the air warmer, it is also drier, since its dew-point has been lowered by removal of some of the water vapour. The warm, dry wind which blows in the lee of the Alps is known as the *föhn*; other large mountain masses have similar winds which can be excessively warm and dry; an example is the *chinook*, blowing to the east of the Rockies. Even the relatively small mountains in Britain give a similar effect though on a smaller scale. Thus, the south shore of the Moray Firth is a favourable *föhn* area when moist south-westerly winds blow over the Scottish Highlands.

6.2 GROWTH OF PRECIPITATION PARTICLES

6.2.1 *Growth by condensation*

Because the water droplets and ice particles in a cloud are denser than the air in which they are suspended, they fall under the influence of gravity. Their fall-speeds are small—a few centimetres per second at the most (see Table V)—so that either they evaporate in the unsaturated air below the cloud base or they are prevented from falling, relative to the ground, by upcurrents of air within the cloud. To reach the ground then, the cloud particles must grow into faster-falling precipitation particles whose sizes are enormous in comparison. An idea of the necessary change in size may be seen by noting that a raindrop with a radius of 1 millimetre contains the same volume of water as 1 million droplets of radius 10 μm, a size commonly found in clouds. It is the purpose of this section to outline the mechanisms which can account for such growth, and in Sections 6.3 and 6.4 we will apply them to particular cloud types.

The first scientific attempt to explain the formation of clouds and rain was made by Hutton in 1784. He ascribed them to the mixing of two moist air masses having different temperatures. Quantitatively, this was shown later to be impossible because such a process can give only very small amounts of water (see Section 3.3.1).

In the late nineteenth century, when both the effects of adiabatic expansion and the need for condensation nuclei were first understood, it was thought that simple condensation alone could produce particles of the size commonly found in precipitation. Certainly the concentration of water liberated by adiabatic expansion of saturated air can be far greater than that obtained by mixing, but a difficulty arises from the huge concentrations of condensation nuclei normally present in the atmosphere—often in the order of 1000 per cubic centimetre. Such large numbers of nuclei mean that correspondingly large numbers of droplets will form which must therefore be small in size and the time to grow to raindrop size becomes an important factor. Laboratory experiments show that, while a droplet can grow from a radius of 1 to 10 μm in a minute or two (10–100 s), to increase the radius by a further factor of 10 will take half an hour to 2 hours (1000–10 000 s) and to double it again may take up to eight hours (30 000 s). During this period of growth the fall-speed of the particle will be increasing up to about 1–2 m s^{-1} (see Table V, p. 93). Hence, while some particles may stay in the cloud long enough to grow to drizzle size, it is extremely unlikely that any particle can attain anything like raindrop size before falling out of the cloud into unsaturated air.

6.2.2 Growth by the Bergeron process

The fact that condensation alone is too slow to produce precipitation-sized particles prompted a search for other mechanisms which would be able to do so. In 1933 the Norwegian meteorologist Bergeron pointed out the significance of the coexistence of both supercooled droplets *and* ice particles in a cloud whose temperature is below 0 °C. In such a cloud the air will tend to remain saturated with respect to water because droplets are present, but at the same time it will be *super*saturated with respect to ice (see Section 3.1.4 and Table I) so that water vapour will be deposited on to the ice particles which will then grow into crystals. The removal of the water vapour required for crystal growth makes the air unsaturated with respect to water so the droplets evaporate. This process continues until either the droplets are completely evaporated or the growing crystal becomes so large that it falls from the droplet cloud. It can be seen that the crystals grow at the expense of the droplets. If the concentration of crystals is small, say, one crystal per thousand or even per million droplets, then it is possible for each of them to grow relatively very large—even several millimetres across. Not only can this process give large particles but it is also rapid—large crystals forming in a time of the order of 10 to 30 minutes. Once formed, these larger crystals fall into lower levels of the atmosphere where, if the temperature is above 0 °C, they melt to water drops, perhaps reaching the ground. As early as 1921 Guilbert had suggested, from an empirical point of view, that ice was essential to the formation of rain in clouds, but it was left to Bergeron to give a theoretical explanation. The mechanism is known as the *Bergeron process*, or sometimes as the Bergeron–Findeisen process through the contribution of the latter in Germany in 1938.

After its introduction, the Bergeron theory was widely accepted as an explanation of the formation of all precipitation of any appreciable intensity. There is, in fact, much evidence to support it. For example:

(a) Aircraft observations of precipitating clouds show that many of them extend upwards to levels where the temperature is below 0 °C, that is, to levels where the presence of ice particles is not only possible, but where samples have also been taken which proved their existence.

(b) Observation from the ground shows a clear relation between the transformation of *Cu* into *Cb* (when ice particles first become numerous in the cloud tops) and the development of showers. This can be seen easily by eye, especially in winter.

(c) Quite small *Cu* in winter can change into *Cb* and give a shower whereas much larger *Cu* in summer remain unchanged. This is because the level in the atmosphere at which temperatures are low enough to promote the formation of ice is much lower in winter.

The sizes and concentration of ice particles forming in a supercooled cloud depend upon the temperature of the cloud, because the first ice particles to appear are the result of freezing of supercooled droplets by contact with freezing nuclei whose activity depends upon the temperature. If the cloud is too warm (between 0 °C and about −10 °C) the concentration of ice particles is very small because there are very few nuclei active in this temperature range and so each particle can grow very large. On the other hand, if the cloud is too cold (below about −30 °C) the concentration of ice particles becomes so large that none of them can grow to any appreciable size at the expense of the remaining droplets because there are too many particles competing for the limited amount of water available. The Bergeron process is favoured by clouds which have:

(a) temperatures in the range −10 to −30 °C (these values are only approximate; they vary greatly with time and place because of the wide variations in the nature and concentrations of natural freezing nuclei);

(b) small liquid water contents (otherwise another mechanism becomes important—see Section 6.2.6).

6.2.3 *Growth by collision*

In spite of the undoubted reality of the Bergeron process, it can explain only the formation of large ice *crystals* (and water drops produced by their melting). The formation of other precipitation types remains unexplained, and another mechanism is needed, especially as there is much evidence to show that some precipitating clouds have temperatures wholly above 0 °C throughout their lives and in these clouds precipitation must therefore develop in some way not involving the ice phase. Such 'warm clouds' were first noticed particularly in tropical and subtropical regions, but they are now known to be fairly common also in temperate latitudes, including Britain (see, for example, Section 6.4.2). A clue to the nature of this new mechanism can be found by closely examining the structures of precipitation particles. For example, a *snowflake* is seen to be composed of a jumble of crystals loosely held together, that is, it is an *aggregate of crystals*. Also, a *snow pellet* is seen to be nothing more than a pellet of rime, a collection of frozen cloud droplets loosely attached to one another and caught by the pellet as it falls through the cloud. This is just the way by which rime accumulates on a plant when a supercooled fog blows past it (see Section 4.2.2); a snow pellet is an *accretion of rime*. Hence, it is seen that both snowflakes and snow pellets seem to be formed by the joining of the relatively small cloud particles to give the much larger precipitation particles—either *crystals* to give *snowflakes*, or *droplets* frozen together to give *snow pellets*. Extending this idea, there seems no reason why droplets should not coalesce to form larger, liquid drops.

Before considering the problem of how cloud particles come to join it is interesting to note some of the methods which have been used to measure the sizes and concentrations of droplets and crystals in clouds:

(a) By exposing a microscope slide coated with a film of magnesium oxide or oil to the air for a fraction of a second and then photographing it. The particles are either permanently caught in the oil or they leave imprints in the magnesium oxide after evaporation.

(b) By exposing a thin aluminium foil which becomes dented by the impact of the particles. Measurements by methods (a) and (b) are made from aircraft flying rapidly through the cloud.

(c) Indirectly, by using measurements of optical phenomena, for example, coronae and fog-bows.

We must now look for a method whereby the cloud particles are made to join. Several have been suggested, for example:

(a) Since the particles in a cloud occur over a wide range of sizes (its spectrum of sizes) they fall at different rates, the largest falling fastest, so that a relatively large particle can sweep up the smaller ones lying in its path.

(b) Attraction between particles having opposite electrical charges.

(c) Attraction between particles having similar sizes caused by a reduction of pressure between them when they fall side by side. (A similar phenomenon is shown by two boats moving side by side at the same speed.)

Of these the first seems to be the most important; the second may be of value on some occasions but little is known about it; all other methods so far suggested seem to be unimportant.

The nature of precipitation formed by collisions depends upon the types of cloud particles present. Thus, if they are:

(a) wholly *liquid*, then *rain* or *drizzle* is formed, and the process is known as *coalescence*;

(b) wholly *crystalline*, then *snowflakes* are formed, and the process is known as *aggregation*;

(c) a *mixture* of droplets and ice particles, then snow *grains*, snow *pellets* or *hail* are formed, and the process is known as *accretion*.

To understand the conditions in a cloud which favour the growth of precipitation by a collison mechanism, we shall consider coalescence first in some detail.

6.2.4 *Growth by coalescence*

Consider a cloud composed wholly of water droplets with a *spectrum of radii* ranging from 5 to 30 μm, say. The origin of this range of sizes may be explained by a corresponding range in sizes of the condensation nuclei upon which they form. The largest droplets are present in very small concentrations compared with the smaller ones, and they fall with greater speeds. This may be explained by considering the forces acting upon a droplet falling freely through the air— its weight and atmospheric drag, a quantity whose magnitude increases with the fall-speed. When the two forces just balance the droplet has no acceleration so it falls with a constant speed known as its *terminal speed*. The larger its weight, the larger the drag needed to balance it and hence the larger the terminal speed. Table V gives values of terminal speeds for particles in which we shall be interested. A large droplet falling through a cloud of smaller ones sweeps up and combines with a large fraction of them lying in its path. The final size to which a droplet can grow in this way clearly must depend upon its rate of growth and the length of time it spends within the cloud.

First consider the rate of growth. It depends upon: the efficiency with which the droplet sweeps up its smaller neighbours, the amount of water in the volume swept out, and the fall-speed of the drop.

TABLE V. *Typical terminal speeds of precipitation particles*

	Water drops			
	Radius		Terminal speed	
	μm	mm	cm s^{-1}	m s^{-1}
Cloud droplets	1	0·001	0·01	
	5	0·005	0·25	
	10	0·01	1·0	
	50	0·05	25	
Drizzle	100	0·1	70	
	250	0·25		2·1
Raindrops	500	0·5		4·0
	1000	1·0		6·5
	1500	1·5		8·1
	2000	2·0		8·8
	2500	2·5		9·1

	Snowflakes	Graupel
Diameter	Terminal speed	
mm	m s^{-1}	
2	1·2	1·5
3	1·4	2·0
4	1·6	2·3
5	1·7	2·5
10 to 40	1·0 to 1·7	approx. 2·7

(a) The *efficiency*, E, may be expressed by the ratio

$$E = \frac{\text{mass of water in droplets combining}}{\text{mass of water in droplets in the volume swept out}},$$

where the volume swept out is the cross-sectional area of the 'sweeping-droplet' multiplied by the path length. E is known as the *collection efficiency*; if all the droplets in the volume are swept up then $E = 1$. The value of E depends upon both the fraction of droplets which coalesce on collision and the difference in size between the 'sweeping-droplet' and those swept up. It seems that most droplets do combine on collision. The importance of the difference in size was first pointed out in 1948 by Langmuir in the United States of America, and his theoretical calculations have since been verified by direct observation. He showed that only those droplets with radii greater than about 20 μm have any appreciable value of E. We may note here that E can be greater than one because droplets just outside the volume being swept out may be drawn into the wake of the larger one and so combine with it by approaching it from above.

(b) A cloud with a large *liquid water content*, W (the amount of liquid water, in grams, suspended in one cubic metre of air), contains a large amount of water in the volume being swept out and is therefore most favourable to the growth of precipitation. The value of W varies within wide limits. It is greater (1) in summer than in winter (because vapour pressures are higher, and there is more water vapour available for condensation); (2) in low-level clouds rather than others (for the same reason); and (3) in cumuliform clouds rather than stratiform clouds—especially the tops of cumuliform clouds whose air has been cooled considerably since it first became saturated whilst rising.

(c) The faster a droplet falls, the faster it sweeps up its neighbours and the larger it becomes. The resulting increase in terminal speed causes even greater rates of growth, so the droplet gets larger at an ever increasing rate.

By considering these three quantities, E, W and terminal speed, we see that growth of a droplet by coalescence is an accelerating process. In fact, most of the time required to produce a raindrop of radius 2 millimetres, say (found to be from about 20 minutes to 1 hour), is used up in forming an intermediate drop of radius 0·5 millimetres, and the last stages occur very quickly. For this reason we can speak of the time required to form a raindrop without specifying its exact size.

Even though the rate of growth may be sufficient to produce a raindrop, that drop may not form if the length of its life within the cloud is too short. This period depends upon:

(a) The *fall-speed* of the particle relative to the cloud (not relative to the ground, since the cloud may contain upcurrents of air capable of suspending the drop within the cloud. The drop must have a terminal speed through the air which exceeds the upcurrent or it will not fall to the ground).

(b) *Depth* of the cloud, for a drop will fall out of a shallow cloud faster than from a deep one.

(c) *Lifetime* of the cloud, for even if all the other conditions are fulfilled, a cloud which lasts less than about 20 minutes will be incapable of producing rain in this way.

Summarizing, although the factors controlling growth of precipitation by coalescence are many and complex, we can see that the following are favourable to it:

(a) Broad spectra of drop sizes, with some radii greater than about 20 μm.

(b) Large liquid water contents (say, at least about 1 g m^{-3}).

(c) Moderate updraughts (say, 1 to 5 m s^{-1}).

(d) Deep clouds (say, several thousand metres).

(e) Persistent clouds (life, say, at least half an hour).

These results will be used in Sections 6.3 and 6.4.

6.2.5 *Growth by aggregation*

Aggregation of ice crystals to form snowflakes is a process similar to the coalescence of droplets. Crystals are likely to interlock on collision because of their complex shapes, particularly the dendritic forms (see Figure 21 and Section 3.4.2). In 1958 Hosler put forward evidence from experiments to show that sticking occurs mostly by ice-to-ice cohesion resulting from the existence of water-like films on the surfaces of crystals, especially when they are growing in an atmosphere supersaturated with respect to ice. When two such surfaces come into contact the films unite and the water, finding itself with ice on both sides, freezes so joining the ice surfaces. The tendency to unite decreases as both the temperature and the degree of supersaturation decrease, so that aggregation occurs most readily at temperatures between 0 °C and −4 °C. This suggests that, in a cloud of falling flakes, aggregation to give large flakes occurs preferably in this temperature range. In support of this, it is an observed fact that snowflake sizes generally decrease with temperature. In an ice crystal cloud which is slowly dissipating by evaporation, for example, *Ci*, *Cs* or *As*, no aggregation occurs.

It has been found that aggregation is most likely either in those clouds which are composed largely or wholly of ice crystals (for example, some forms of *As* and *Ns* and the older, uppermost parts of *Cb*) or in supercooled clouds of small liquid water content which are capable of growing the necessary crystals by the Bergeron process (see Section 6.2.2).

6.2.6 *Growth by accretion*

Careful examination of a snowflake will usually show the presence of a few tiny, frozen cloud droplets adhering to the crystals, having been swept up as the flake fell through a supercooled cloud. On striking the flake, a supercooled droplet freezes almost instantaneously and retains its approximately spherical shape. If a falling flake acquires only a few frozen droplets, a *rimed snowflake* is produced, but if accretion is more pronounced other types of particles appear— *graupel* and *hail*. Each of these can form only in a cloud containing *both* super-cooled droplets and ice particles, and since this is a cloud of the same type which favours the Bergeron process, then either accretion or the Bergeron process can operate. The liquid water content largely controls which is dominant: a *small* liquid water content favours growth by *deposition* of water vapour on to ice crystals since the droplets present are so widely spaced that they are not easily swept up; a *large* liquid water content favours growth by *accretion* of droplets on to the ice particles because the droplets are more crowded.

Two types of ice accretion can be deposited on a falling particle:

(a) *Rime*, an accumulation of discrete frozen droplets with many air spaces, resulting in an ice of low density (about one-tenth that of water) which is white and opaque. The particle is soft and fragile and takes the form of *graupel* (*snow grains* or *snow pellets*). For rime to form, the liquid water content must not be too large (say, $0 \cdot 1$ to $1 \cdot 0$ g m^{-3}). Typical clouds from which graupel reaches the ground are: thick *Sc str op* in winter (snow grains) and *Cu con* or *Cb cal* in winter (snow pellets). Graupel may form at higher levels in summer, but it melts before reaching the ground.

(b) *Clear ice*, with little entrapped air, resulting in ice of greater density (about three-quarters that of water) which is hard and not easily broken. *Hail* is formed in this way in clouds with a large liquid water content (say, greater than $1 \cdot 0$ g m^{-3}). The reason for the compact nature of the ice is that the supercooled droplets are collected so rapidly they have time to unite before freezing. More will be said about hail in Section 6.4.3.

6.2.7 *Sizes of precipitation particles*

We all know that raindrops vary greatly in size, the largest being found, for instance, in a summer thunderstorm. Now these drops have a maximum size, which can be demonstrated quite easily by observing the breakup of a thin stream of water issuing from a tap, or by noting the size of the largest drops which form slowly by dripping from the tap. In 1904 Lenard first measured their diameter, finding them to be about $5 \cdot 5$ millimetres. The reason for this maximum is both interesting and important; it depends upon the change in shape of a falling drop as its size increases. Studies by Blanchard in 1950 and McDonald in 1954 have shown that falling drops are not 'tear-shaped', but approximately spherical as long as they are small (radius less than about 0·5 millimetre), whilst larger drops have flattened bases, giving a 'bun-shaped' vertical cross-section. A large drop is so much flattened (even going as far as a hollow lower surface) that it becomes unstable, so that a small disturbance, such as may be produced by collision with another drop or by turbulence in the atmosphere, results in its breakup. These changes of shape with size also account for the fairly constant terminal speed of the largest drops (Table V).

There does not appear to be any theoretical reason to expect a maximum size for snowflakes or snow pellets, but in practice there does seem to be a rough upper limit to flake sizes at about 50 millimetres and to pellets at about 5 millimetres, but on rare occasions they are larger. The largest flakes are usually seen with temperatures just above 0 °C when they become slightly wet and adhere to each other readily. The size of hailstones will be considered in Section 6.4.3.

6.2.8 *Chain reactions*

All the mechanisms of precipitation growth discussed so far cause a reduction in the number of cloud particles as a result of the preferential growth of a few. It is possible, however, for new particles to form by the breaking of larger ones, the fragments themselves then growing until they in turn break up. In this way one particle can lead to the formation of perhaps hundreds or even thousands of others, as long as there is sufficient water vapour available to feed them and enough time for growth. Such a process is a *chain reaction* and several have been suggested:

(a) In 1948 Langmuir pointed out that in those clouds which are very favourable to coalescence the drops may grow so large as to become unstable and break up. Most of the fragments are still sufficiently large to grow rapidly, reaching maximum size again inside a few minutes, only to be followed by a repetition of the process. We would expect drop-breaking to be most likely in the huge towering *Cu* and *Cb* of summer.

(b) In 1950 Ludlam suggested that when hail grows in a cloud of large liquid water content the droplets may be swept up so rapidly that they form a liquid film before freezing, and some of this water may be shed from the larger stones as they fall rapidly through the cloud (see also Section 6.4.3).

(c) Because of their very delicate and complex structures, growing ice crystals easily shed minute splinters of ice which can then act as further centres for the deposition of water vapour.

(d) When a water drop starts to freeze, an ice-film often forms over its surface. Now freezing water expands, since the density of ice is less than the density of water, so that when the interior of the drop freezes it ruptures the outer ice-film which then sheds small fragments. This process has been observed in the laboratory with both rain-sized drops and the much smaller super-cooled cloud droplets.

6.2.9 *Changes in precipitation whilst falling between cloud and ground*

So far in Section 6.2 we have studied how precipitation forms within the cloud; before it can be observed at the ground it is often greatly modified. Some of the changes which may occur are:

(a) *Evaporation* in the unsaturated air below the cloud. Precipitation seldom reaches the ground if the cloud base is higher than about 3000 metres because this depth of clear air is usually sufficient to evaporate even the largest particles. Evaporation first removes the smallest particles and therefore increases the proportion of the largest. An example of this selective evaporation can be seen by comparing rain falling from *As*, which has a relatively high cloud base and gives mostly moderate-sized drops, with rain from *Ns*, which has a lower base and gives drops with a wide spectrum of sizes. Again, this idea can be used when a few large drops are seen to fall from a sky covered with *St*—such drops will not have originated in the *St*, but will have fallen through it from some other cloud layer above. Light snow falling from *As* may sublime below the cloud base and the resulting cooling of the air may be sufficient to cause instability so that the air descends in pockets, producing a diffuse mottling to the undersurface of the cloud, resembling feeble mamma. It may be seen on a developing layer of *As* whose precipitation has not yet been able to reach the ground.

(b) Snow and hail start to *melt* when they fall below the 0 °C level or, more exactly, when the aspirated wet-bulb temperature of the surrounding air is above 0 °C. Because of their large terminal speeds and high densities, hailstones may fall several kilometres before completely melting; large stones can reach the ground even when the 0 °C level is as high as four kilometres. Snowflakes melt more quickly, usually within the first 500 to 1000 metres below the 0 °C level, so that snow is rarely observed at the ground if the screen temperature exceeds about 4 °C, but showers of snow are sometimes seen when the temperature is a high as 7 °C as long as the air is dry, that is, as long as the wet-bulb temperature does not much exceed 0 °C. The melting of continuous snow, say, from *Ns*, causes a progressive cooling of the atmosphere below the 0 °C level because latent heat is used up, until a deep layer, perhaps 1000 metres deep, may become nearly isothermal at 0 °C. In this way, the 0 °C level may build downwards so far that snow, which earlier had melted on its way down, can eventually reach the ground unmelted. It has been known for widespread rain to be replaced by snow in this way.

6.3 PRECIPITATION FROM STRATIFORM CLOUDS

6.3.1 *General*

Stratiform clouds form by condensation processes which are slow but persistent, and the rate at which precipitation falls from them approximately balances the rate at which water vapour condenses, so the precipitation is steady but of small intensity compared with showers. The ideas developed in Section 6.2 will now be used to discuss the types of precipitation associated with different stratiform clouds.

6.3.2 *Stratus and stratocumulus*

St and *Sc* are shallow (perhaps 1000 metres deep), they have weak updraughts and small water contents. For most of the year they are warmer than about −10 °C and are unlikely to contain ice particles. Precipitation grows by coalescence, so we can see from Section 6.2.4 that it will be light and composed of small drops, that is, drizzle, or at the most, fine rain. The added cooling accompanying lifting over high ground can intensify the precipitation to a thick, fine rain of small drops, such as is described by the phrase 'Scotch mist'.

In winter, when the cloud tops may have a temperature at or a little below −10 °C, a few ice particles may develop. If the cloud is not too dense, these can grow by the Bergeron process into isolated well-developed crystals which can be observed easily by eye, especially if they are allowed to settle on dark paper or clothes. With a denser cloud, accretion occurs, giving small pellets of rime known as snow grains or granular snow, sometimes as fine as table salt. Always the rate of precipitation is very small.

6.3.3 *Thin layered clouds*

Sc, Ac and *Cc* of the species *str* and *len* are often shallow, say up to 300 metres deep, and they have very small liquid water contents. Any precipitation from them is rare and largely confined to a few occasions when the cloud temperature is below about −20 °C—cold enough to allow the local growth of ice particles in the form of small flakes or pellets and in sufficient numbers to fall out as definite streaks, or virga.

6.3.4 *Thick layered and multi-layered clouds*

Much of our widespread and prolonged precipitation falls from these clouds. Their upper parts usually extend to levels where temperatures are low enough to produce numerous crystals and flakes, which melt after falling below the 0 °C level to form drops whose growth may subsequently continue by coalescence with water droplets in clouds at lower levels. When the cloud is deep the precipitation may be moderate or even heavy in intensity. In multi-layered clouds the uppermost and coldest layer can act as a generating level for snowflakes which, on falling into any lower, supercooled layer, soon grow by the Bergeron process. For example, Figure 36 shows crystals falling from *Cs* into a lower layer of supercooled *Ac* where they grow rapidly and aggregate to snowflakes. Such growth of precipitation in one cloud layer after falling from a higher layer is known as *seeding* (see also Section 6.5).

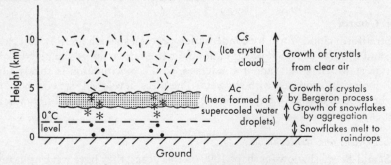

FIGURE 36. *An example of growth of precipitation in multi-layered clouds*
Crystals falling from *Cs* seed a lower layer of supercooled *Ac*.

6.3.5 *Unusual types of precipitation*

On a few occasions in winter rain may be seen reaching the ground in a super-cooled state. Such *freezing rain* (or *drizzle*) is found when it falls into a layer of air near the ground whose temperature is below 0 °C (Figure 37). It is most frequently found when thick stratiform clouds, resulting from frontal ascent or convergence, move across a land surface which has been intensely cooled in winter. Because the temperature of the ground and the air in contact with it is below 0 °C, the drops freeze after striking the ground but they do so slowly

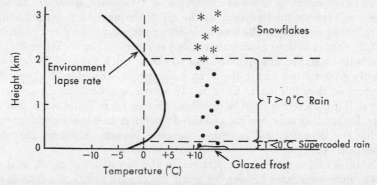

FIGURE 37. *A situation favourable to the formation of freezing rain and glazed frost*

enough for them to join first into a film of water. In this way a sheet of clear ice, or *glazed frost*, covers the ground, buildings, trees, etc. In severe cases overhead power cables may become so thickly coated as to break under the increased weight. Fortunately such severe glazed frosts are rare in Britain.

On a very few occasions the drops may have been frozen on their way down to give *ice pellets* or *frozen rain* (*drizzle*). They are irregular in shape and usually show evidence of remains of snowflakes from which the original drops formed by partial melting. (More spherical ice pellets are formed by the freezing of raindrops and are considered in Section 6.4.3 as a form of hail).

6.4 PRECIPITATION FROM CUMULIFORM CLOUDS (SHOWERS)

6.4.1 *General*

Cumuliform clouds form by rapid and localized ascent of air, and the rate at which precipitation falls from them cannot balance the rate at which water vapour condenses, at least during the early stages, so the water at first accumulates within the cloud and the ensuing rate of precipitation is heavy compared with that from stratiform clouds. It takes the form of showers, sometimes known as 'thundery rain' if it falls from medium cloud in an unstable atmosphere.

6.4.2 *Showers from cumulus clouds*

Inside a large *Cu* cloud rain-sized drops can form by coalescence in about 20 minutes. If the updraughts are weak, say, one metre per second, then a summer cloud with a minimum depth of only about 1500 metres is found to be capable of producing them. Slight showers of fine rain (that is, small drops) are sometimes observed from such clouds, particularly on windward coasts and over high ground. Most *Cu* clouds, however, have stronger updraughts, in the order of five metres per second (10 knots), and they need to be at least 3000 metres deep to produce raindrops, some of which may be of maximum size. These very large raindrops may then break up into several large and many small drops most of which may well have fall velocities small enough to ensure that they are recycled within the cloud. The larger ones may again grow to the maximum size as part of a chain reaction for the multiplication of drops of raindrop size. Gradually, however, the weight of water will reduce the buoyancy of the air in the upcurrent, so reducing the upcurrent that not only is the supply of water vapour for further condensation cut off, but the raindrops are released to fall as a heavy shower. When *Cu* clouds spread below a stable atmospheric layer to give a thick sheet of *Sc (Ac) cugen*, the partly grown drops within them can no longer be suspended by the reduced upcurrents so they fall out as fine rain. In winter, cumuliform clouds must be deeper still if coalescence is to be possible, because their liquid water content is smaller. In fact, they must become so much deeper that their tops ascend to levels so cold that the Bergeron and accretion processes become dominant and the *Cu* changes into *Cb*.

The initial formation of precipitation in the top of a *Cu* cloud, as distinct from *Cb*, may sometimes be seen by eye if the cloud growth proceeds as far as the minimum depth needed but no further. When such a cloud decays by mixing with its environment the smaller cloud droplets evaporate first, leaving a tenuous cloud of larger particles which persists for a minute or more. In 1956 Ludlam introduced the term *fibrillation* to describe this process. It is the formation of a fleeting streakiness which either evaporates into the drier surroundings or sinks into the top of a new, rising *Cu* tower where the drops can start to grow again.

6.4.3 *Showers from cumulonimbus; the formation of hail*

When *Cu* tops reach levels where the temperature is below −10 °C a few of the freezing nuclei become active and some of the liquid droplets freeze to become ice particles. If the cloud continues to grow and the temperature of the tops continues to fall more and more freezing nuclei become activated and when the temperature falls to about −40 °C the cloud becomes glaciated and no supercooled water droplets are present. In summer a *Cu* cloud which has a temperature inside its top as low as −10 °C is likely to be at least 3000 metres deep and so will contain much liquid water, including large drops grown by coalescence. At low temperatures (−20 °C) small cloud droplets freeze rapidly on coalescing with ice particles producing an aggregate of white opaque grains of ice–snow pellets. At temperatures only just below 0 °C, relatively large droplets striking a snow pellet tend to spread over the surface before freezing and so produce a layer of clear ice with little entrapped air.

When hailstones are examined it is found that they may have a core of soft opaque ice (a snow pellet, Figure 38(*b*)) or of clear transparent ice (a frozen raindrop, Figure 38(*a*)). This core may be surrounded by alternate layers of clear and opaque ice usually up to five in number though many more layers may be present on occasion (Figure 38(*c*)). We can imagine that a particle grows at an increasing rate by sweeping up supercooled droplets in an updraught until it reaches such a size that its fall-speed exceeds the updraught and it falls out of the cloud as a

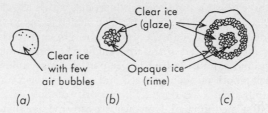

FIGURE 38. *Cross-sections through some typical ice pellets and hailstones*
(a) Ice pellet (frozen raindrop). (b) Ice pellet, or small hail. (c) Multi-layered hailstone.

relatively large hailstone. The layers of opaque and clear ice can be explained as being acquired in different parts of the cloud within the temperature range 0 to –40 °C—opaque in regions of low temperature and water content, clear ice in regions of high temperature and water content. Although this explanation of the formation and constitution of hailstones is plausible, when the idea is examined more closely various difficulties appear; for example the very large updraught required to support the heaviest hailstones is likely to carry the growing hailstone above the –40 °C level where growth then ceases.

The argument above tacitly assumes that the updraught is vertical and approximately constant with height. When we take into account the fact that the horizontal wind often changes with height so that the updraught may be appreciably tilted and also that its speed may increase with height, these difficulties largely disappear. Consider Figure 39 which represents a model of a cumulonimbus producing hail.

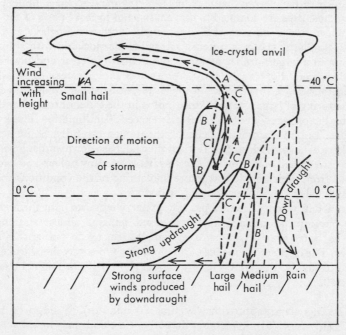

FIGURE 39. *The structure of a hailstorm*

In the diagram three possible trajectories of particles within the cloud are labelled A . . . . . A, B . . . . . B, and C . . . . . C. The first one represents the path of particles rising once only through the updraught, growing to small hail pellets and being swept forward ahead of the updraught where they may fall out, melt and reach the ground as rain. Rather larger particles may follow such paths as B . . . . . B, growing both on their upward and on their downward journeys and re-entering the updraught lower down. The majority of these will escape from the updraught and fall as medium hail. However, we can envisage that a small proportion of these particles re-entering the updraught are of just the right size to be lifted slowly in an updraught increasing in speed with height and to grow at such a rate that their fall-speed remains just a little below updraught speed until they reach the level of maximum updraught. They are then carried out of the updraught and fall along trajectory C . . . . . C to reach the ground as large hail. The duration of a fall of hail may vary from a few seconds to about half an hour, though most frequently lasting about five minutes. To be effective this process required considerable vertical development in the cloud, vigorous updraughts and a plentiful supply of moisture.

In winter, the transformation of *Cu* to *Cb* takes place in clouds which are more shallow and have smaller water contents than in summer. Hail formation is therefore unlikely so that, initially, the precipitation usually takes the form of *snow pellets*. As rapid glaciation proceeds, accompanied by a decrease in the concentration of liquid water, aggregation of crystals to *snowflakes* takes place. In a well-developed winter shower both types can be observed. In *very* cold weather, *Cb* clouds may be very shallow so that glaciation occurs when they may be only 1000 metres deep or even less. The result is a rapid transformation of the *Cu* into a mass of falling snowflakes.

Our study of hail shows that it is most likely inland in the summer; indeed, our worst hailstorms are confined to that season and to those parts of the country where intense convection is most likely, namely south-east England and the east Midlands. Similarly, the most intense rainfalls are associated with summer *Cb*. Hail occurs in a large number of forms other than the clear and layered types already discussed. For example, we sometimes find clusters of small stones frozen together, or ice shells with liquid centres, or irregular lumps of ice formed by the shattering of larger stones. Most hail is in the order of 6 to 12 millimetres in diameter but on rare occasions stones may exceed 50 millimetres. The structures of hailstones can be examined by cutting them open carefully. Snow pellets are either spherical or somewhat conical when they fall base downwards. Forms intermediate between hail and snow pellets and between snow pellets and snowflakes occur, and from their detailed structure something of the conditions inside the cloud which led to their formation can be deduced.

If, on a day when *Cb* develops, the upper winds are light throughout the unstable part of the atmosphere, each cloud remains almost stationary and deposits its water over a comparatively small area, up to perhaps 250 square kilometres. The complete sequence of precipitation types may then be observed in one place. Roughly, the life-cycle of a model shower cloud might be as shown in Figure 40.

(a) *Growing* stage (approximately first 20 minutes). Buildup of towering *Cu* (Figure 40(*a*)). Remember that other *Cu* clouds may have been present earlier but they failed to develop further.

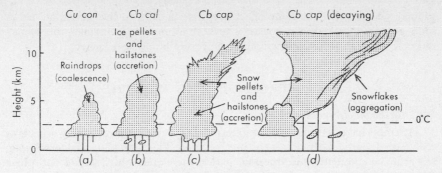

FIGURE 40. *Life-cycle of a model shower from Cb in summer*

Type of shower at ground: (a) Large raindrops. (b) Large raindrops and/or hailstones. (c) heavy rain and/or hail. (d) moderate-sized raindrops.

(b) *Mature* stage (approximately a further 20 minutes). Heavy rain, by coalescence or by melting of hail (some of which may reach the ground). In this period the complete change *Cu con* → *Cb cal* → *Cb cap* takes place (Figure 40(*b*) and (*c*)). Thundery activity may develop during this stage—see Chapter 7.

(c) *Decaying* stage (next half to two hours). Rain gradually decreases in intensity; most of it (especially later) has formed by the melting of snowflakes (Figure 40(*d*)). Any thundery activity gradually weakens.

Although this complete sequence may be observed sometimes, winds are usually sufficient to cause at least a slow drift of the clouds so that a series of showers may be experienced at any one place, each at a different stage of development. Heavy downpours are likely to indicate an early stage. It should be remembered that most shower clouds consist of groups of *Cb* (see Sections 5.4.3 and 7.3.1). If wind shear is present, the glaciated upper parts of the *Cb* may either stream out ahead (if horizontal wind increases upwards) or trail behind (if wind decreases upwards) and in either situation there is a greater tendency for convection to continue longer than would occur if the shear were absent. In the former event, the first precipitation noted by an observer falls from the old cloud top and takes the form of snowflakes (or rain, by their melting), followed by either snow pellets in winter or heavy rain and hail in summer, all coming from the more active part of the cloud. In the latter event when the anvil trails behind, the sequence is reversed.

6.4.4 'Thundery rain'

Thick masses of *Ac flo* and *Ac cas*, which form when the atmosphere is unstable through a deep layer in the middle troposphere, can also develop precipitation by processes similar to those operating inside *Cu* and *Cb*. The instability at medium levels is reflected in the sporadic nature of the rainfall at the ground, often called 'outbreaks of thundery rain'. Degeneration of the cloud into masses of precipitation transforms it into *As* or *Ns*. This type of weather is most common over southern Britain in summer; the rain may be heavy and prolonged at times, and hail or thunder may occur.

Rain is occasionally found falling from a cloudless sky. This occurs when the original cloud has completely decayed before all its rain has reached the ground. Incidentally, the same effect may also be seen to the immediate lee of high ground or windward coasts where showers are occurring—the clouds disperse whilst drifting downwind but the precipitation lasts longer.

6.5 CONTROL OF THE WEATHER

6.5.1 *Artificial clearance of clouds*

Attempts have been made during the past 30 years to control certain aspects of the weather, particularly clouds and precipitation. The methods used depend upon the artificial stimulation of processes within the clouds which would not otherwise occur naturally because of some deficiency. Foremost among these have been attempts at 'rainmaking', and it is their importance which has spurred on much of the modern research into the physics of clouds and precipitation.

A convenient way to clear a sheet of cloud is to induce the droplets to grow and fall out as precipitation. If the cloud is supercooled but not so cold that ice particles have formed naturally, then by inducing some of the droplets to freeze they will grow by the Bergeron process and fall out. However, if too few ice particles form, they leave behind most of the water droplets; if too many form, none can grow large enough to fall out and an ice crystal cloud remains. We can see that a cloud suitable for dispersal in this way must be supercooled, but not colder than about -10 °C. The droplets may be induced to freeze in two ways: firstly, by adding lumps of solid carbon dioxide, 'dry ice', which cools some of the droplets to below -40 C, that is, below the temperature at which they freeze spontaneously (see Section 3.4.1); secondly, by adding *artificial freezing nuclei* which are active at temperatures higher than those at which natural nuclei are active. In 1947 Vonnegut found that silver iodide particles acted as very efficient artificial freezing nuclei: when put into a supercooled cloud they produce ice particles as long as the temperature is below about -5 °C. Solid carbon dioxide is best dropped from an aircraft in the form of pellets. Silver iodide is best prepared as a 'smoke' by burning a solution made from it and sodium iodide in acetone. After seeding artificially in this way, precipitation often develops within about a half-hour, falls out, and leaves a clear lane along the line of seeding. The results can be spectacular but may not be permanent because cloud can subsequently re-form in the clear space. It must be noted that the cloud types suitable for clearance are very restricted—they must be supercooled stratiform clouds which are not precipitating naturally.

6.5.2 *Artificial stimulation of precipitation*

In the more arid parts of the world an increase in rainfall would be a great boon, and this has led many workers to investigate methods whereby greater rainfall might be induced artificially. If a cloud is to give rainfall of any intensity it must have a large liquid water content and so must be of a cumuliform type. Most attention has been given to those *Cu* which can be induced to give showers in situations when they would not do so naturally. Both methods of seeding described above have been used, and although precipitation undoubtedly follows many occasions of seeding, it is usually difficult to tell whether the precipitation has been produced by the seeding or by natural processes. This is because those clouds which have properties just right for seeding are usually found in a mixed population of clouds, some of which give precipitation naturally. Because of this the results are difficult to assess, and the infrequent occurrence of suitable clouds makes the chance of giving any appreciable increase in precipitation over a wide area unlikely.

In clouds whose temperatures are wholly greater than 0 °C, seeding has been carried out by putting a spray of large droplets into the cloud base, or by using a fine powder of salt as a source of large condensation nuclei.

The results of seeding widespread layer cloud are even more difficult to determine. In Britain, experiments were made from Salisbury Plain in the 1950s using generators for silver iodide smoke, but two difficulties were found: the slowness with which the smoke diffused upwards to levels where it became effective (say 3000 to 6000 metres); and the decrease in activity of silver iodide whilst it was exposed to sunlight.

An interesting method which may prove of value economically in favoured areas is the seeding of extensive orographic clouds over wide mountain ranges. If the seeding is done at the upwind end of such a cloud and if it is supercooled, prolonged snowfall may be induced on the hills during the winter. When the snow melts, it adds to the natural water supply for hydroelectric and irrigation schemes. Experiments on these lines have been made in the United States of America, Sweden and Australia.

Ideas upon the possibility of forming rain in clouds go back to the nineteenth century. In 1891 Gathmann in the United States of America suggested the use of liquid carbon dioxide, and in 1904 Gregory in Australia suggested liquid air. No results seem to have been recorded, probably because of the impracticability of the experiments at that time.

Although much time and money have been spent in rainmaking experiments, on the whole returns have been meagre. Certainly many early claims of success were far too optimistic and many workers in this field have been slow to realize the necessity for valid statistical tests of their claims.

6.5.3 *Some further attempts at weather modification*

Some success has been achieved in the clearance of fog over a very limited area such as an airfield runway. Supercooled water fogs have been cleared by inducing glaciation. Although not economically viable in this country, the method has been used for example in Scandinavia. During the Second World War fog was cleared temporarily from runways by the use of FIDO (Fog Investigation Dispersal Operation), a system developed in Great Britain. This involved the burning of petrol at intervals along the runway to raise the air temperature and is completely uneconomic except in wartime. Jet engines have also been used.

In some parts of the world hailstorms are a serious problem, causing extensive damage to crops and property. Attempts have been made to seed growing Cb with silver iodide taken into the cloud by rockets or shells from anti-aircraft guns. By these means it is hoped to increase the concentration of effective freezing nuclei and prevent the development of large hailstones by encouraging the growth of a large number of small stones. Some success has been claimed for this technique, but recent models of hail-producing clouds suggest that the location of seeding is of crucial importance; seeding certain regions of the cloud might have the effect of encouraging the growth of large hail.

A similar technique has been applied to the problem of reducing the destructive power of hurricanes (see Section 10.4.2). The cumulus clouds farther from the centre than the cloud surrounding the eye are seeded to encourage development to the outflow level, so that ascent of air takes place over an increased area instead of being concentrated in the eye wall. This has the effect of reducing maximum wind speeds in the region just outside the eye. The technique is still in the research stage—to date only four hurricanes have been seeded. Results so far have been promising and it is hoped to extend experiments to the western North Pacific over an area to the east of the Philippines in 1977 and 1978. Progress must be gradual as it is essential to ensure that interfering with a hurricane does nothing to increase its strength or to alter the speed or direction of motion.

BIBLIOGRAPHY

DELDERFIELD, E.R.; 1953. The Lynmouth flood disaster. Exmouth, Rayleigh Press.

HARDMAN, M.E.; 1968. The Wiltshire hailstorm, 13 July 1967. *Weather*, **23**, pp. 404–415.

HOLLAND, D.J.; 1967. The Cardington rainfall experiment. *Met Mag*, **96**, pp. 193–202.

JACKSON, M.C.; 1975. Annual duration of rainfall intensity. *Met Mag*, **104**, pp. 243–248.

KESSLER, E.; 1973. On the artificial increase of precipitation. *Weather*, **28**, pp. 188–194.

LOWNDES, C.A.S.; 1971. Substantial snowfalls over the United Kingdom, 1954–69. *Met Mag*, **100**, pp. 193–207.

LUDLAM, F.H.; 1956. The structure of rainclouds. *Weather*, **11**, pp. 187–196.

McFARLANE, D. and SMITH, C.G.; 1968. Remarkable rainfall in Oxford. *Met Mag*, **97**, pp. 235–245.

MANLEY, G.; 1969. Snowfall in Britain over the past 300 years. *Weather*, **24**, pp. 428–437.

MARWITZ, J.D.; 1973. Hailstorms and hail suppression techniques in the U.S.S.R., 1972. *Bull Amer Met Soc*, **54**, pp. 317–325.

MASON, B.J.; 1971. The physics of clouds, 2nd edition. Oxford, Clarendon Press.

MASON, B.J.; 1975. Clouds, rain and rainmaking, 2nd edition. Cambridge, University Press.

ODDIE, B.C.V.; 1959. First results of the Meteorological Office experiments on the artificial stimulation of rainfall. *Met Mag*, **88**, pp. 129–135.

PEDGLEY, D.E.; 1970. Heavy rainfalls over Snowdonia. *Weather*, **25**, pp. 340–350.

PRIOR, M.J.; 1975. The heavy rainfall over northern England in July 1973. *Met Mag*, **104**, pp. 108–118.

SALTER, PAULINE M. and RICHARDS, C.J.; 1974. A memorable rainfall event over southern England. Parts I and II. *Met Mag*, **103**, pp. 255–268 and 288–300.

SIMPSON, JOANNE; 1967. An experimental approach to cumulus clouds and hurricanes. *Weather*, **22**, pp. 95–114.

CHAPTER 7

THUNDERSTORMS

7.1 INTRODUCTION

7.1.1 *Some definitions*

Whenever thunder is heard a *thunderstorm* is said to be taking place. Such storms probably give the most spectacular weather observed over Britain. A wide range of different phenomena accompanies a thunderstorm, but the physical processes involved in its life-cycle often take place quickly enough to be easily followed by a single observer. Indeed, the changes sometimes take place so rapidly and are so complex that it is difficult to keep pace with them.

Observations suggest that thunderstorms develop mainly in those clouds which contain large concentrations of liquid water *and* ice coexisting at temperatures well below −20 °C. From the discussions in Sections 6.2.4 and 6.4.3 we see that the only clouds which have these properties are those cumuliform clouds whose tops extend to at least 6 kilometres, the most common being *Cb*. Most thunderstorms develop in *Cb* clouds but some occur in thick unstable medium cloud. It follows, then, that thunderstorms occur when conditions are favourable for the development of heavy showers and, indeed, many of the properties of a storm are of the same type as found with showers—the difference is merely one of intensity on most occasions. But thunderstorms show some features which are unique, including of course thunder and lightning. Also, our most violent weather in the form of intense hailstorms, squalls and tornadoes are almost always found with thunderstorms. In this chapter we will be concerned particularly with these special phenomena; Sections 5.4.3 and 6.4.3 should be consulted for the general properties of *Cb* clouds and the formation of precipitation within them.

Lightning is a discharge of static electricity—a lightning flash is simply an enormous spark. The discharge may take place either between a cloud and the ground or between two clouds or, rarely, from the cloud into some other part of itself or the clear air around it. In Britain about 40 per cent of lightning flashes are of the first type. There is a popular distinction made between 'forked' and 'sheet' lightning, but it is not real: the first is merely lightning whose actual path of discharge is visible as an irregular, highly luminous spark, whilst the second is a diffuse glow which is all that can be seen when the discharge path is obscured by cloud or precipitation. If there is no obstruction between a storm and the observer, lightning at night may be seen at a great distance, even exceeding 150 kilometres.

Thunder is the sound produced by the violent expansion of the air as it is suddenly and intensely heated along the path of the flash. The rumbling is caused by differences in the times required for sound to reach the observer's ear from various parts of a flash, which is sometimes 2 kilometres long; echoes add to it. Since the light from the flash is received at the eye almost instantaneously whereas sound waves travel at only about 1 kilometre in 3 seconds,

107

there is always a time lag between the lightning and its accompanying thunder, and the greater the distance of the flash from the observer the greater is this lag. A flash at 1 kilometre gives a lag of 3 seconds, so timing the lag in seconds and dividing the result by 3 gives the distance of the flash in kilometres. Thunder can be heard fairly often up to 15 kilometres from its source but only rarely above 30 kilometres.

7.2 LIGHTNING

7.2.1 *Electrical charge generation*

The first experiments to show that lightning is an electrical phenomenon were made in 1752 by Franklin in the United States of America. Since then much work has been done, notably by British meteorologists and physicists, in attempts to explain its formation. In 1909 Simpson put forward an explanation of the generation of charge within a thundercloud based on the observation that when a water drop is violently disrupted the largest fragments become positive and the fine spray from the surface film is negative. Now the largest drops fall fastest whilst the spray is taken aloft in the upcurrents of the *Cb*, so the cloud near its base should develop a positive charge whilst its top becomes negative. However, Simpson's suggested *spray mechanism* proved to be unlikely when the actual distribution of charge within a cloud was measured, indirectly by Wilson in 1916, and directly by Simpson and Scrase in 1937. They found that the polarity of the cloud was in fact the reverse, with positive charges aloft centred at about 6 kilometres, and negative below, at about 3 kilometres (Figure 41). However, a small positive centre was often found at about 1·5 kilometres, usually associated with heavy rain. Although the spray process may be used to explain this lower positive charge at least in part, we cannot use it to explain the generation of the main charges in the cloud.

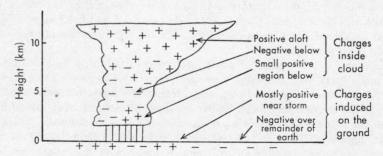

FIGURE 41. *Distribution of electrostatic charge in and near a model Cb cloud*

The presence of the *ice phase* in the upper parts of a developing thundercloud has been considered fundamental to the generation of electrostatic charge. As evidence of this it has been stated that:

(a) The main charge generation occurs at temperatures well below 0 °C.
(b) There is a marked association between the development of intense electric fields within the cloud and visible glaciation.
(c) No storms have definitely been observed to start in all-water clouds.

However, although these statements are probably true for middle and high latitude clouds, they may not constitute the whole truth since evidence is coming to light that the electrification of warm clouds does occur in low latitudes. If this evidence is confirmed, clearly other processes not involving the ice phase must be operating in charge generation.

In 1948, before this latest evidence was available, Workman and Reynolds in the United States suggested that the *freezing of water* might be important in the generation of charge. Observations have shown that the growth of rime in a super-cooled cloud is accompanied by a pronounced separation of charge such that the rime becomes negative whilst minute splinters of ice, shed during the freezing of the drops (see Section 6.2.8), become positive. In a *Cb* where extensive formation of snow pellets is taking place, a riming mechanism is believed to be important in the generation of thunderstorm electricity. It is complicated by the effects of temperature differences between the various particles thrown together within the cloud, and also by impurities normally dissolved in the droplets. Riming, however, seems capable not only of producing charge quickly enough to maintain the normal frequency of flashes (about one per 20 seconds in the storm's most active stage) but also of giving the correct polarity to the cloud, because the negative precipitation falls out whilst the positive charge is carried aloft on the minute ice splinters. It cannot be said that any of the numerous other mechanisms which have been suggested fully satisfy these conditions, but the electrical processes taking place inside a thunderstorm are so complex that an understanding of them is far from complete.

It is probable that several charging processes operate simultaneously, and that the dominant one may alter during the life of any one storm. Of other charging processes which have been put forward, Wilson's *induction mechanism*, first proposed in 1916, is interesting. It relies upon the prior existence of both an electric field between a negative earth and a positive ionosphere, and also a mixture of positive and negative ions in the air. This situation is a normal one, the ions being produced by radioactivity (near the ground) and by cosmic radiation (especially at high levels). A water drop or ice particle suspended in such a field has charges induced upon it—positive below and negative above. When it falls through the field at a speed greater than that of the ions (which is about 1 centimetre per second) it will tend to sweep the ions up, but will gather only those of sign opposite to that of the sweeping surface. Hence, negative ions are swept up and so the particle becomes negative. Experiments have confirmed this mechanism. It cannot, however, initiate the bulk of the charge separation but may become important during the later stages of the storm when the electric fields and concentration of ions are larger. Other mechanisms have been suggested, most of which involve the ice phase; at the same time a coherent theory for the electrification of warm clouds has yet to be found.

7.2.2 *Electrical discharges and currents*

The lightning *discharge process* is extremely rapid but it has been studied by two methods:

(a) Photography, especially by Schönland since 1938, using a rotating camera, invented in 1902 by Boys, which spreads the image over the film and gives a time resolution of about 1×10^{-6} second, or 1 *microsecond* (μs).

(b) Measuring changes in the electrostatic field accompanying changes in charge distribution during a flash.

Numerous observations show that a lightning flash consists of several successive strokes following the discharge channel made by the first stroke. These strokes, which are shown diagrammatically in Figure 42, may be divided into:

(a) A *leader stroke*, either continuous and building steadily downwards from cloud to ground at 15 km s⁻¹ or more, or a more rapid, intermittent stroke, progressing in steps of about 100 metres over periods of about 2 μs with 'resting' intervals of about 50 μs (that is, overall speeds of about 200 km s⁻¹). It is usually branched and reaches the ground in about 0·01 s. Only the stepped leader is luminous and then but faintly.

(b) A *return stroke*, continuous from ground to cloud and lasting about 40 μs during which it is intensely luminous. It has high speeds of 20 000 to 150 000 km s⁻¹, with a maximum current of about 20 kiloamperes.

(c) After about 0·1 s a *dart leader* may descend along the same path in about 0·001 s, to be followed by a second return stroke. Dart and return strokes may be repeated several times, during which successive portions of the negative charge in the lower part of the cloud are 'tapped'.

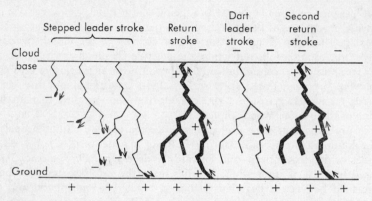

FIGURE 42. *Successive stages during the life of a lightning flash*

A flash occurs when sufficient charge has been generated, and separated, to overcome the resistance of the air. Dry air is a very poor conductor of electricity and very large field strengths are needed to break down its insulating power—about 3×10^6 V m⁻¹—but the presence of water drops in the air reduces the maximum required. An overall voltage in the order of 10^8 volts will give a flash. Most discharges bring down negative charge from the cloud, averaging 20 coulombs. Assuming an estimate by Brooks of 100 flashes per second over the whole earth's surface and of these about 10 per cent striking the ground, we see that an upward current of about 200 amperes (A) (that is, $20 \times 100/10$ coulombs per second) is being continuously maintained by the lightning from all thunderstorms in the world. (We are taking here the convention that a current flows from positive to negative centres.) However, an even larger current is produced by another process associated with thunderstorms. This is a 'point discharge' from objects on the ground such as vegetation and buildings. Streams of positive ions which were originally induced on the ground below the negative cloud move upwards in the intense electric fields between ground and cloud giving a current density of about 0·02 A km⁻² or, say, 0·5 A per storm. For the whole world this gives about 1000 amperes, continuously supplied by the 1800 storms estimated to be existing at any one time. Upward currents of a similar magnitude have been measured from aircraft above the tops of storms and they probably exist also in association with other precipitation areas which do not develop sufficient charge to give lightning. Precipitation carries a current too, and in the opposite direction, but it is small.

From these results we see that an upward current in the order of 1000 amperes is maintained by thunderstorms. It is balanced by downward ionization currents in the fair weather areas, so we may picture the lower atmosphere as being a poorly conducting medium lying between the earth and the ionosphere which act as the plates of a leaky condenser (Figure 43).

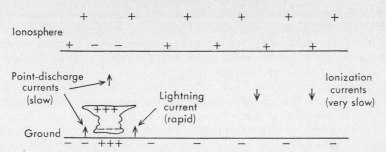

FIGURE 43. *Diagrammatic representation of electric currents in the atmosphere*
The arrows indicate direction of currents, that is, direction of movement of positive
charges.

The rapid fluctuations in current strength during a lightning flash result in the generation
of radiation detectable on radio receivers and known as atmospherics. The form and frequencies of this radiation vary with the nature of the discharge and with its distance from
the observer, and may be used to locate its source with fair accuracy for distances up to
2500 kilometres. Simultaneous bearings are taken from several observing stations scattered
over Britain, using *cathode-ray direction-finding* (CRDF) equipment.

The effects of lightning striking the ground are well known, and are often attributed
to 'thunderbolts' as though they were solid objects hitting the ground at high speeds.
However, many of the effects are really the result of intense heating as the current passes
through the ground, which may be fused locally into contorted masses. Water may boil
almost instantly, giving rise to such phenomena as the shattering of masonry and the
splitting of trees ('blasted oaks'); both are caused by the explosive expansion of the water
as it boils. Estimates of lightning frequency in Britain are about one flash to ground per
square kilometre per year, so it can be easily seen that the chances of a person's being struck
are very small.

7.2.3 *Unusual lightning*

Apart from normal lightning discharges, there are some others of interest:

(a) *Ball lightning.* A roughly spherical mass of glowing air usually about 100 to
200 millimetres in diameter which persists and drifts with the wind. It is
very rare and its real existence has been doubted, but there have been some
well authenticated reports. Its cause is unknown, but it may be a mass of
air which has been energized in the path of a lightning flash and then
slowly loses its energy. On striking an earthed object it seems to disappear,
thus giving the impression of passing through the object.

(b) *St Elmo's fire*, or *corposants*. A more or less continuous, luminous electrical
discharge in the atmosphere, emanating from objects at the earth's
surface such as ships' rigging, wind vanes, lightning conductors, and
even hair and pencils. It is a point discharge under the influence of
intense electric fields such as may be found near thunderstorms or in
snow blowing along the ground.

(c) *Flachenblitz* (*Rocket lightning*). A rare form which extends upwards from a
Cb top and ends in clear air.

7.3 THUNDERSTORM WEATHER

7.3.1 *Downdraughts and cells*

Between 1946 and 1949 an extensive investigation into the properties of
thunderstorms was made in the United States of America—the 'Project Thunder-

storm'—and its results were published in 1949 by Byers and Braham. A great deal of new information was obtained about the structure and life-cycles of thunderstorms, and the discussions on the development of *Cu* and *Cb* clouds in Sections 5.4 and 6.4 are to some extent based on their findings. Now we will consider some further results of special interest.

When the precipitation within a developing *Cb* has accumulated and starts to fall out, a *downdraught* is initiated, caused by the drag exerted by the enormous numbers of precipitation particles as they fall through the air. This downdraught has a temperature below that of its surroundings (both cloudy air, and clear air around the cloud) because the descending air warms at the saturated adiabatic lapse rate whereas the lapse rate in the environment is somewhat greater than the saturated adiabatic lapse rate, and so is the lapse rate in the rising cloudy air (at least near the edges) as a result of mixing with cloud-free air. The lower temperature of the downdraught maintains the downward convection, so it continues to descend and on striking the ground fans out beneath the *Cb* cloud, more particularly in the direction of the cloud's motion (Figure 44). Eventually it may extend for some tens of kilometres from the centre of the storm. When first formed the downdraught may coexist with adjacent powerful updraughts, but as the cloud ages and much of the precipitation falls out, most of the cloud is occupied by downdraughts, so little new condensation is possible and the precipitation intensity therefore gradually dies away. Similar downdraughts are associated with showers which have not developed into thunderstorms, but then they are usually weaker.

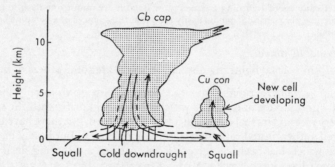

FIGURE 44. *Cold downdraught spreading below a vigorous Cb cloud*

A spreading cold downdraught passing an observer is indicated by pronounced changes in the weather:

(a) *Wind*. The onset is sudden, sometimes violent, and is shown by the rapid replacement of a wind initially flowing gently towards the developing *Cb* by a squally wind, with gusts perhaps exceeding 60 knots, blowing outwards from the centre. The greater is the distance of the observer from the storm's centre or the older the cloud, the weaker is the squall. These features are best seen when the prevailing wind is light because if it is strong it interferes with the downdraught, producing complex effects.

(b) *Temperature*. The downdraught is cold, so its onset is marked by a sudden fall in temperature, changing by perhaps 10 °C in as many minutes on extreme occasions.

(c) *Precipitation.* In its early stages, when the downdraught edges are still beneath the *Cb* cloud, the start of rain and squall are roughly coincidental, but later, when the cold air has spread well away from the precipitation area, the squall is not accompanied by precipitation.

(d) *Pressure.* Since the downdraught is cold we would expect its greater density to produce a sudden pressure rise. This is often observed and may amount to several millibars, so that in thundery weather the barogram shows a very irregular trace. Where the downdraughts from several adjacent storms combine a small anticyclone may form, some tens of kilometres across, with general outflow of air around its edges (this flow is not geostrophic—the pressure gradient force tends to be balanced by friction).

Thunderstorms seldom consist of single *Cb* clouds. Most are irregular groups with each cloud in a different stage of development or decay. A group as a whole may last for hours but individual clouds or *thunderstorm cells*, have a life of half an hour to an hour. The separate nature of the cells can be observed quite easily at night by noting the positions and frequencies of lightning flashes. Each cell produces flashes actively for about 20 minutes, the period usually starting some minutes after the first heavy precipitation reaches the ground (which may not, of course, be near the observer). New clouds form preferably near the advancing edges of the spreading downdraughts (Figure 44), particularly where two downdraughts are in contact.

Widespread outbreaks of thunderstorms sometimes develop in deep unstable medium clouds (see Sections 5.4.4 and 6.4.4). Such storms are usually less severe but may be spectacular in their lightning displays, the flashes being visible over a greater length because of the high cloud base (sometimes above 3000 metres). They are a feature of summer weather in southern Britain.

7.3.2 *Multicell and supercell storms*

More recently research workers have made intensive studies of individual storms by using multiple radars and aircraft observations and several categories of hailstorm have been proposed. Of these categories perhaps two of the most basic are the multicell and supercell storms and these will be described briefly. The essential difference between these two types is that a supercell storm is dominated by a single cell which attains a quasi-stationary structure whereas a multicell storm consists of a sequence of evolving cells each of which may go through a life-cycle resembling that described by Byers and Braham.

Most hailstorms fall into the multicell category. According to the model such a storm consists of several cells (cumulus clouds) at different stages of development at any one time, the lifetime of an individual cell being about 45 minutes. Figure 45 depicts a vertical section along the storm's direction of travel through a model of a hailstorm which occurred over the State of Colorado in the United States in 1973. The solid lines represent streamlines of flow relative to the moving system (broken lines represent flow out of the plane of the section). The open circles denote the trajectory of a hailstone during its growth from a droplet at the cloud base. Light stippling denotes the extent of the cloud while darker stippling represents intensity of radar reflectivity. The horizontal line N–S through the section represents the path of an aircraft which obtained the profile of vertical velocity plotted (smoothed) at the foot of Figure 45. The diagram can be interpreted in two ways: either as an

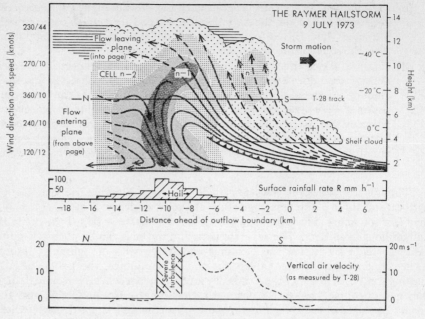

FIGURE 45. *Vertical section through a multicell storm*

instantaneous view of a typical storm made up of four individual cells labelled $n+1$, n, $n-1$, $n-2$ or as showing four stages in the evolution of a single cell. Thus $n+1$ represents a discrete new growing cumulus cloud just about to enter the storm, n represents the stage at which the first echo was recorded, $n-1$ the mature stage and $n-2$ the decaying stage. In the diagram the airflow in each cell has been drawn relative to the individual cell which is itself moving through the storm as a whole. This accounts for the apparent anomaly of the trajectory of the growing hailstone crossing over the streamlines. This model adds detail to the earlier one and, in particular, suggests that each cell entering the storm does not feed the mature hail cloud but itself becomes the main hail cloud in due course.

The term supercell was first coined by Browning in 1962 to describe the quasi-stationary state attained by the Wokingham storm of July 1959, which produced hail with some stones over an inch in diameter along a path about 200 kilometres long across south-east England. Hail was falling from the storm almost continuously for 4 hours. Supercell storms can develop from multicell storms; when they do they usually turn to the right. To develop, such storms require a rather special set of environmental conditions, including strong instability in combination with strong vertical wind shear and usually strong vertical wind veer. They are characterized by a single persistent cell which generally travels to the right of the mean wind (in the northern hemisphere) while retaining a quasi-steady structure with the updraught and downdraught coexisting for periods which are long compared to the time taken for air to pass through the storm. Moist air enters the storm from the right flank at low levels, turns as it ascends and leaves at high levels from the front of the storm on the left flank. Cold dry air at middle levels enters the storm from the right, crosses ahead of the updraught and descends to the ground on the left flank within the rain area.

The model described below was developed from close study of a supercell storm over Colorado in June 1972 which produced serious hail damage over a path of about 200 kilometres. Only a few of the more important features will be mentioned here. Figure 46 represents a section orientated along the direction of travel of the storm (south-south-east) and shows the outline of the cloud, the stippled area being echo free while the intensity of radar reflectivity is indicated by two densities of hatching. The thin lines are streamlines of the airflow relative to the storm, while the short thin arrows skirting the boundary of the vault represent a hailstone trajectory. The weak echo vault shows the region of intense updraught where particles can ascend from cloud base to the −40 °C level in as little as 5 minutes, so restricting their growth that the radar echo is weak. The forward overhang or embryo curtain occurs mainly above the inflow, ahead of and partly round the weak echo vault. It is a region where particles are able to recycle and grow into hailstones.

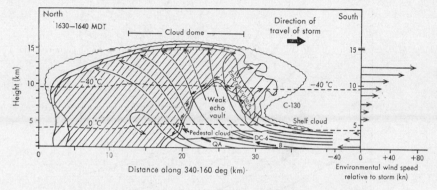

FIGURE 46. *Vertical section through a supercell storm*

Bold arrows denote wind vectors in the plane of the diagram as measured by aircraft represented by QA and B.

In Figure 47 hatching again indicates regions of radar echo while the dotted circle represents the extent of intense updraught in the middle troposphere. The arrowed lines are streamlines of airflow relative to the storm, showing:

(a) environmental flow at middle levels diverted round the main updraught, and

(b) low-level inflow towards the updraught (dashed lines) with the corresponding high-level outflow.

The embryo curtain lies in the region where the outflow is shown above the inflow to the south-east and south of the vault.

Typical hailstone trajectories are shown in vertical section in Figure 48(*a*) and in plan view in Figure 48(*b*). These trajectories are consistent with the airflow shown in Figures 46 and 47. Four trajectories are sketched. Trajectory 0 represents the path of cloud particles which are drawn into the main updraught and ascend to above the −40 °C level so quickly that they have no time to grow larger than a few hundred micrometres before they are carried off into the anvil and take no part in the formation of hail. Trajectory 1 represents the path of particles rising more slowly near the edge of the main updraught and hence with more time to grow to millimetric size. Most of these will be carried away in the northern branch

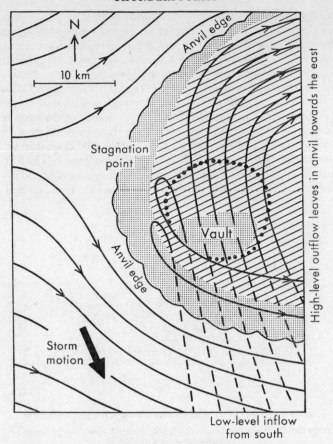

Storm
motion

Low-level inflow
from south

FIGURE 47. *Plan view showing the principal features of the airflow within and around a supercell storm*

of the high-level outflow (Figure 47) but some will follow the southern branch and may be large enough to fall along trajectory 2 in the weak updraught of the embryo curtain and reach the lowest part of the curtain where they may re-enter the foot of the main updraught. Here many are likely to encounter intense updraughts and follow the dotted trajectory (Figure 48) but some will follow trajectory 3 in air with a large water content where there is a chance of their terminal fall-speed being matched closely with the updraught velocity. These are the particles which may grow rapidly as they cross the vault (Figure 46) and fall along the northern boundary as damaging hail (the hail cascade).

This supercell model has been outlined as it applied to the Colorado storm of June 1972 which moved from north-north-west to south-south-east. It should be noted that in the United Kingdom supercell storms such as the Wokingham storm of July 1959 move almost invariably from south-west to north-east.

7.3.3 *Tornadoes*

A tornado is a violent whirlwind with an approximately vertical axis and with a diameter of about 50–200 metres. An essential requirement is a strong updraught

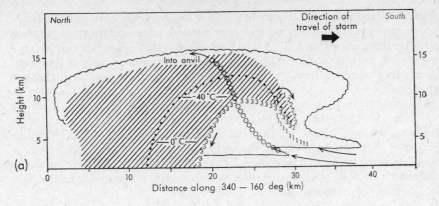

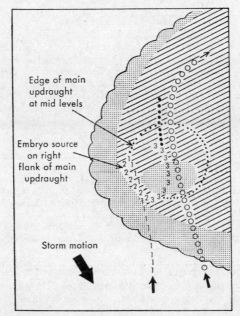

FIGURE 48. *Schematic model of hailstone trajectories within a supercell storm*

    (a) Vertical section along the direction of travel of the storm.
    (b) Plan view.

and hence it occurs in association with thunderstorms. The most violent tornadoes occur in the central plains of the United States where they can cause extensive damage to property and loss of life. In this region dry air from the Rocky Mountains or from the Mexican Plateau frequently flows over warm moist surface air from the Gulf of Mexico thus providing a suitable environment for the development of supercell storms (see Section 7.3.2) with their associated intense convection and it is here that tornadoes may form, usually on the right flank of the storm. They can sometimes be identified on the radar screen by the presence of a marked hook echo. Winds of about 200 knots or more may occur near the tornado centre in an extremely narrow ring; at the actual centre the horizontal wind is zero but the updraught is capable of lifting heavy objects. This rapid rotation requires a very

large horizontal pressure gradient to produce the necessary inward acceleration
(see Section 2.4.2) and there may be a pressure fall of several hundred millibars
at the centre of the tornado. It is the suction effect of this pressure deficit which is
responsible for severe structural damage by the outward collapse of buildings.
Damage can also result from the twisting effect of the large horizontal wind shears
which arise.

Fortunately tornadoes in the United Kingdom are much less severe than the
typical tornado of the American Mid West, although severe structural damage has
been caused to buildings on occasion and loss of life has been recorded. The average
frequency appears to be about one a month. As in America the typical tornado in
Britain is associated with warm moist air at the surface moving from a southerly
point and it usually travels from a direction between south and west on the right
flank of a storm accompanied by hail and thunder.

BIBLIOGRAPHY

BROWNING, K.A.; 1962. Cellular structure of convective storms. *Met Mag*, **91**, pp. 341–350.
BROWNING, K.A.; 1967. The growth environment of hailstones. *Met Mag*, **96**, pp. 202–211.
BROWNING, K.A.; 1968. The organization of severe local storms. *Weather*, **23**, pp. 429–434.
BROWNING, K.A. and FOOTE, G.B.; 1976. Airflow and hail growth in supercell storms and
    some implications for hail suppression. *Q J R Met Soc*, **102**, pp. 499–533.
GOLDIE, E.C.W. and HEIGHES, J.M.; 1972. Investigation of a United States Midwest
    tornado. *Met Mag*, **101**, pp. 270–278.
HARDY, R.N.; 1971. The Cyprus waterspouts and tornadoes of 22 December 1969.
    Met Mag, **100**, pp. 74–82.
KAMBUROVA, PETIA L. and LUDLAM, F.H.; 1966. Rainfall evaporation in thunderstorm
    downdraughts. *Q J R Met Soc*, **92**, pp. 510–518.
LACY, R.E.; 1968. Tornadoes in Britain during 1963–66. *Weather*, **23**, pp. 116–124.
LAMB, H.H.; 1957. Tornadoes in England May 21, 1950. *Geophys Mem*, **12**, No. 99.
LUDLAM, F.H.; 1963. Severe local storms: a review. *Met Monogr, Boston*, **5**, No. 27, pp. 1–30.
MARWITZ, J.D.; 1972. The structure and motion of severe hailstorms. Part I: Supercell
    storms; Part II: Multi-cell storms. *J Appl Met*, **11**, pp. 166–179 and 180–188.
MASON, B.J. and MOORE, C.B.; 1976. Theories of thunderstorm electrification. *Q J R Met
    Soc*, **102**, pp. 219–240.
ROACH, W.T.; 1967. On the nature of the summit area of severe storms in Oklahoma.
    Q J R Met Soc, **93**, pp. 318–336.
ROACH, W.T.; 1968. The Barnacle tornado. *Weather*, **23**, pp. 418–423.
STANFORD, J.L.; 1975. A pictorial survey of Iowa tornadoes over three-quarters of a century.
    Weather, **30**, pp. 43–55.
WHYTE, K.W.; 1974. The tornadoes of 26 June 1973. *Met Mag*, **103**, pp. 160–171.
WRIGHT, P.B.; 1973. A tornado in south Yorkshire and other tornadoes in Britain.
    Weather, **28**, pp. 416–428.

CHAPTER 8

OPTICAL PHENOMENA

8.1 IN ICE-CRYSTAL CLOUDS

8.1.1 *Simple haloes*

When the sun shines through *Cs* or *Ci* clouds a ring of bright light may sometimes be seen with the sun lying at its centre. It is usually reddish on the inside and white on the outside, and is known as a *halo*. The cause of this phenomenon is *refraction* of the sun's light by the ice crystals composing the cloud. A crystal's edge acts as a refracting edge in the same way as does the edge of a glass prism.

The deviation, D, or change in direction of the light after passing through a prism, depends upon the way the light passes through the prism, but its magnitude is never less than a minimum value, D_0, given by the equation:

$$\sin \frac{D_0 + A}{2} = \eta \sin \frac{A}{2},$$

where A is the refracting angle and η is the refractive index. Minimum deviation occurs when the plane containing the incoming and outgoing light beams is perpendicular to the refracting edge, that is, when the light passes through the prism symmetrically (Figure 49(a)). In Section 3.4.2 we saw that the typical shape of an ice crystal is the hexagonal

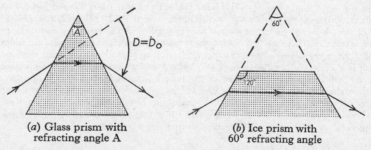

(a) Glass prism with refracting angle A

(b) Ice prism with 60° refracting angle

FIGURE 49. *Light passing symmetrically through a prism at minimum deviation*

prism whose sides are inclined to each other at 120°. Refraction cannot occur in an ice prism with a refracting angle of 120°, but it can take place between alternate sides because they intersect at 60° (Figure 49(b)). Taking $\eta = 1\cdot31$ for ice, $D_0 = 22°$ approximately. Refraction can also occur between a side and an end of a crystal intersecting at 90°, and for such an edge, $D_0 = 46°$ approximately.

Cs often consists of multitudes of small hexagonal prisms or spatial groups of prisms, which are oriented randomly in the air. When parallel light from the sun passes through them, some of the refracting edges will be lying in positions such that they can refract the light into the observer's eye. For example, A in Figure 50(a) is such a crystal. Let the direction of any particular beam of refracted light, measured relative to a line joining the eye and the sun, be R. By geometry, $R = D$. Now, of those edges refracting light into the eye, some will do so with minimum deviation (for example, A in Figure 50(b)) and the light

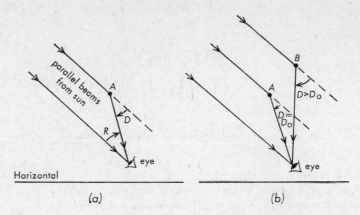

FIGURE 50. *Refraction of light into the eye to produce a halo*

(a) Refraction by a randomly orientated crystal, A.
(b) Refraction by two crystals—A refracting at minimum deviation and B at greater than minimum deviation.

from them will have $R = D_0$; the remainder (for example, B in Figure 50(b)) will have $D > D_0$ and hence $R > D_0$. No refracted light can enter the eye where $R < D_0$.

The intensity of the refracted light can be shown to be at a maximum where $R = D_0$ and it decreases rapidly as R increases, so rapidly, in fact, that when R exceeds about $(D + 1)°$ it is too small to add noticeably to the general illumination from the sky. The result is to produce an annular image of the sun of angular radius D_0, measured from the sun's centre to the inside of the ring. The most common refracting edges are of 60° which thus produce a halo of radius 22°—the 22° *halo*; much less common is the 46° *halo* formed from the 90° edges.

In taking $\eta = 1·31$ for ice, we neglected the fact that η decreases slightly as the wavelength of the light increases—$\eta = 1·317$ for violet light and $1·307$ for red light, hence the value of D_0 varies slightly with the wavelength. For the 22° halo, $D_0 = 21°34'$ for red, and $22°22'$ for violet light—hence the red halo is smaller and lies on the inside; the outside, however, is not violet because the violet light from crystals refracting at minimum deviation is mixed with other colours from crystals refracting at just over minimum deviation and the result is a white outer edge to the halo. Yellow and green may sometimes be seen just outside the red. Since the central area of the halo has no refracted light it appears rather darker than the sky outside the halo. The 46° halo is coloured similarly but usually less brilliantly although with better colour separation. Numerous other haloes have been observed, with radii from about 7° upwards, but they are formed by refracting edges involving very unusual pyramidal crystals, so they are very rare.

8.1.2 *Other halo phenomena*

So far, we have considered the refraction of light by *randomly* oriented crystals. On some occasions, however, the crystals have a *preferred* orientation depending upon their shape. For example, simple prisms tend to fall with their long axes horizontal, whereas plates fall with their flat bases horizontal; in both instances the largest area possible is exposed to the effects of air drag. Crystals with a preferred orientation produce special refraction phenomena by concentrating light in certain areas outside the corresponding halo, and the shapes of the images vary with the elevation of the sun. Thus, the 60° edges of horizontal prisms give a complicated *circumscribing halo*, seldom complete, being usually represented by short *arcs of contact* to the 22° halo and touching it vertically above or below the sun (upper and lower arcs, respectively). When the sun is low, the upper arc takes the form of a downward pointing *cusp*; with a progressively higher sun, it becomes a *tangent* concave downwards; and finally, when the sun's elevation above the horizon exceeds 55°, the upper and lower arcs may combine to give an approximately *elliptical halo*, with longest axis horizontal, dwindling to become coincidental with the 22° halo when the sun is vertically overhead. If colouring is visible in these arcs, red is again on the sides nearest the sun.

If the 60° edges are vertical, as, for example, in plates or capped columns, then bright spots of light are produced at the same elevation as the sun, one on each side, and just outside the 22° halo; these are *parhelia*, or *mock suns*. When the sun is on the horizon, they lie exactly on the 22° halo but as the sun's elevation increases so they are found farther outside, but they also become less intense, so it is unusual to see a mock sun with an elevation exceeding about 20°; when it is seen, it is still so close to the halo as to seem to lie on it. It is often brightly coloured, again with red inside.

When the 60° edges are vertical, then the 90° edges must be horizontal and refraction through them produces a brightly coloured arc, the *circumzenithal arc*, with its centre directly overhead. Its size decreases as the sun's elevation increases, but it always passes close to the top of the 46° halo so that it is often mistaken as a tangent arc to that halo. A sun's elevation greater than 32° cannot produce this arc.

Combinations of some or all of the halo phenomena so far mentioned are sometimes seen. Figure 51 shows a combination of the commonest forms. Many other refraction phenomena caused by ice crystals have been observed but mostly on rare occasions, and some still remain unexplained.

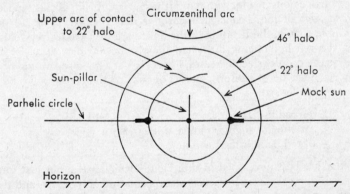

FIGURE 51. *Diagram showing a combination of the commonest halo phenomena*

Ice crystals possess flat faces which can act as mirrors so we would expect some optical phenomena produced by *reflection*. They are sometimes seen, the commonest being:

(a) *Parhelic circle*, from vertical crystal faces. It is a circle whose elevation is everywhere the same as that of the sun, that is, it passes through the sun, and lies parallel to the horizon with its centre vertically overhead. The higher the sun's elevation, the smaller the circle. The parhelia lie on the circle and usually show bright extensions along it pointing away from the sun.

(b) *Sun pillar*, from approximately horizontal crystal faces. It is a bright image of the sun, extending vertically above or below.

Reflection phenomena do not involve the separation of colours so they all have the same colour as the sun, that is, white by day, but reddish near sunrise or sunset.

Halo phenomena may be seen around the moon. Whether around the sun or moon they are seldom complete or persistent because of rapid variations in the composition of the clouds. Not all ice-particle clouds give them, only those containing well-formed crystals—normally those clouds actively developing—and since those are often followed by precipitation well within 48 hours, the occurrence of halo phenomena can be associated with coming precipitation. *Cs* is the best cloud in which to look for halo phenomena, but they may also be seen in other clouds, for example, fallstreak *Ci*, *virga* from *Ac flo*, contrails and anvil tops of *Cb*. Carefully looked for, they can be seen on about one day in two, even if only fleetingly.

8.2 IN WATER-DROPLET CLOUDS

8.2.1 *Coronae and iridescence*

When the sun shines through clouds composed of minute spherical water droplets, each of approximately the same size, refraction and reflection phenomena of the type associated with ice crystals are not observed. Instead, we sometimes see coloured rings of a much smaller radius centred on the sun and having a bluish colour inside, with red farthest away. A ring of this type is known as a *corona*; two or more concentric coronae may coexist. They are produced by *diffraction*, that is, the bending of the light path as it passes very close to the droplets, the red light being diffracted most. The angular radius of a corona increases as the droplet size decreases, hence, to get a good separation of colours, the droplets must be of uniform size. The size of the corona may be used to estimate the droplet size:

Drop radius (μm)	10	5	2	1
Approximate angular radius, in degrees, of red band in first corona . .	2	4	10	20

Coronae are best developed in thin clouds which are either just forming or just dispersing, particularly lenticular clouds and thin *Ac (Sc) str*, and their radii are usually only a few degrees. Only small portions of the larger coronae are normally seen, and since their sizes are very sensitive to the droplet radius, the colours are usually seen in irregular patches and appear especially as pastel shades of pink and green. Patches of cloud coloured in this way are said to show *iridescence*, or *irisation*. Both coronae and iridescence are also produced by the moon.

8.2.2 *Glory*

If you stand with your back to the sun and look down upon the top of a nearby bank of fog or cloud (most readily done in a mountainous region) a corona may be seen around the *antisolar point*, that is, around the point lying on an extension of the line joining the sun to your eye, and near the middle of the shadow of your head. The light which has been diffracted to give this corona is sunlight which has already been reflected inside the droplets of the fog or cloud in such a way that it returns along its original path. Such a corona is known as a *glory*. It is interesting to note that for several people standing side by side, each gives a shadow on the fog but only that of the observer has a glory—the others are unadorned. Each, of course, sees a glory only around the shadow of his own head. The shadow itself is known as a *Brocken Spectre*.

8.3 THE RAINBOW

8.3.1 *The normal rainbow*

A normal well-developed *rainbow* consists of three parts:

(a) *Primary bow*—brightly coloured with violet on the inside, followed by blue, green, yellow, orange and then red on the outside.

(b) *Secondary bow*—lying outside the primary bow and well separated from it. It is less intense and the colours are reversed, that is, red is on the inside.

(c) Several *supernumerary bows*—lying just inside the primary. The colours are rather poorly separated and their intensity is small; only rarely are there more than two.

Conditions fundamental to the seeing of a rainbow are that the observer must stand with his back to the sun and face a mass of falling raindrops illuminated by sunlight. The bows form a set of circular arcs centred at the antisolar point (see Section 8.2.2) and are produced by the refraction and internal reflection of sunlight by the raindrops. The deviation of the light as it passes through a drop is so great as to bend it back almost opposite to its original path.

8.3.2 *The primary bow*

In Figure 52(a) the parallel rays from the sun, A, B and C, each pass through a spherical drop. B and C are refracted twice and internally reflected once so as to emerge as B' and C' respectively, but A is only reflected. Each ray is deviated by a different amount depending upon the initial angle of incidence at the drop's surface, but there is one ray, B, for which the deviation is a minimum, D_0. The light is also dispersed into its constituent colours and just as occurred with refraction by an ice crystal, the violet ray is bent the most; in fact, $D_0 = 139\frac{1}{2}°$ for violet light and $D_0 = 138°$ for red (Figure 52(b)).

Let a ray which is deviated by an angle D so as to enter the eye, make an angle R with a line joining the eye to the antisolar point as in Figure 52(c). By geometry, it is seen that $R = (\pi - D)$. Since the minimum value of D is D_0, the maximum value of R must be $(\pi - D_0)$, that is, refracted light cannot enter the eye from directions having $R > (\pi - D_0)$. The refracted light comes from directions with $R \leqslant (\pi - D_0)$, and it can be shown to have a maximum intensity when $R = (\pi - D_0)$. The combined effect of myriads of raindrops refracting at minimum deviation is to produce an annular image of the sun of angular radius $(\pi - D_0)$, measured from the antisolar point. For violet light $R = 40\frac{1}{2}°$ approximately, and for red 42° approximately, so the image is violet on the side nearest the antisolar point and red on the outside (Figure 52(d)). The sky just outside the bow is relatively dark because no refracted light is received from that region to add to the general background illumination.

8.3.3 *The secondary bow*

Some of the light passing through the drops undergoes two internal reflections, and its minimum deviation is about $233\frac{1}{2}°$ for violet light and about $230\frac{1}{2}°$ for red (Figure 53(a)). Here, $R = (D - \pi)$, so that for minimum deviation of violet light $R = 53\frac{1}{2}°$ approximately, and for red $50\frac{1}{2}°$, giving an annular image of the sun which is violet on the outside and red on the inside (Figure 53(b)). More than two internal reflections can occur but the intensities of

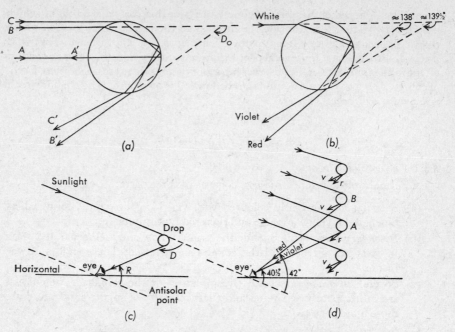

FIGURE 52. *Formation of a primary rainbow*

(a) Illustrating variation of deviation with position of incident beam on drop's surface. Beam *B* shows minimum deviation.

(b) Separation of colours with minimum deviation.

(c) A random drop refracting into the eye.

(d) A set of drops refracting at minimum deviation. For some the light enters the eye — with *A* it is violet, with *B* red.

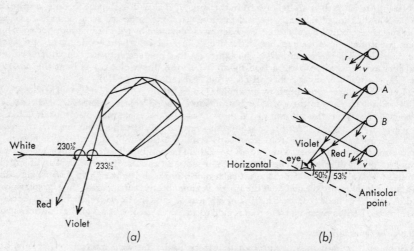

FIGURE 53. *Formation of a secondary rainbow*

(a) Separation of colours with minimum deviation.

(b) A set of drops refracting at minimum deviation. For some the light enters the eye — with *A* it is violet, with *B* red.

the emerging beams are so feeble that bows are very rarely seen. Even the secondary bow is only about one-tenth of the intensity of the primary.

8.3.4 Supernumerary bows

Rays adjacent to B in Figure 52(a) interfere with each other after refraction, giving one or two narrow bows with values of R a little less than $(\pi - D_0)$; their spacing increases as the drop size decreases, so that for raindrops it is less than 1°. If all the drops in the falling rain are of the same size the colour separation is good; usually there is a wide spectrum of sizes giving poor separation and only one supernumerary forms, or, in extreme examples, none because the colours from different supernumeraries overlap to give near-white. The properties of these extra bows can be used to gain some idea of the size and spread of size in the falling raindrops.

8.3.5 Cloudbows and fogbows

When the refracting drops are very small, as in clouds or fogs, the colours of the primary and supernumerary bows are found to overlap, giving an almost white *cloudbow* or *fogbow* (*Ulloa's ring*) which is a little smaller than the normal rainbow—for droplets of radius 5 μm the bow has a radius of about 37°. Any supernumeraries present are also white and well separated—for the 5 μm droplets the separation is about 12°. These bows are very rare, and best seen from an aircraft or mountain, looking down on to fog or cloud.

8.3.6 Unusual rainbows

Primary *lunar rainbows* may be seen sometimes but their colour separation is usually poor. Secondaries are very rare, and so are *lunar fogbows*.

When a rainbow occurs over a calm water surface a *reflected rainbow* may be seen; when complete it is an inverted image of the original. If the flat water surface lies behind the observer, sunlight may be reflected from it before refraction by the raindrops, giving a *reflection rainbow*, the source of whose light appears to be a point with an elevation below the horizon equal to that of the sun above. The bow is an inverted image of the invisible lower part of the original and has its centre above the horizon, that is, when fully formed the arc is more than half a circle. The colours and curvatures of these bows are the same as in the original.

Rainbows may also be seen in waterfall spray and spray from a garden hose. A bow formed in *sea spray* is a little smaller than a normal rainbow because the dissolved salt alters the refracting power of water, and both may be seen co-existing when squally showers are present at sea.

8.3.7 Size of a rainbow

The angular radii of the bows are always as described, but the maximum length of arc visible decreases as the sun's elevation increases, until the sun's elevation exceeds D_0 when no bow is visible above the horizon. It follows from this that rainbows are most likely to be seen in the early morning and late afternoon, and in winter rather than summer. In the middle of a summer day a primary rainbow is impossible. Because of its greater radius a secondary bow is possible over wider time ranges than the primary. A bow seldom extends far below the level of the horizon because, in such a direction, there are too few drops between the eye and the ground to refract light of detectable intensity, unless the observer is on a high place or the drops are closely spaced, when a complete circle is sometimes seen.

Most of our rainbows are found in showery weather and since winds are often from the west when showers occur, a morning bow usually precedes the rain, whereas an afternoon bow follows it. As an observer moves, he sees the rainbow move. This is because after the observer has changed his position the refracted light entering his eye comes from a different set of drops, but still in the same directions relative to the (new) antisolar point. It follows from this, too, that two adjacent observers see different bows. A bow formed by a shower on the horizon is very squat, extending from the ground to cloud base (or to the melting level if the rain starts as solid precipitation at higher levels). A small fragment of a rainbow seen in showery weather is known by mariners as a 'wind-dog'.

8.4 OTHER OPTICAL PHENOMENA

8.4.1 Coloured sky, sun and moon

In Sections 8.1 to 8.3 we have seen how the sun's light may be *reflected*, *refracted* or *diffracted* on its way to our eyes by water drops or ice crystals in the atmosphere. Now the molecules in the atmosphere itself alter the sunlight by a process known as *scattering*, that is, a change in the direction of the light rays as they pass near the molecules. Rayleigh has shown that the extent of scattering decreases as the wavelength increases, so violet light is scattered more than red. Hence, light entering the eye from the sky (as distinct from the sun) will be rich in violet with smaller amounts of the other colours, the result being a *blue sky*. The light from the sun has now been depleted somewhat of its bluer constituents but the loss is small and the sun still appears white. However, near sunrise and sunset the length of atmosphere traversed by a ray entering the eye directly from the sun is much greater than at midday, so the effect of scattering is correspondingly larger, and the sun's light is then distinctly yellowish, or even reddish, because of the much greater loss of its bluer constituents.

When smoke or dust haze is present, the minute solid particles reflect sunlight diffusely, often with no selection of colours, to give a milkiness superimposed on the blue sky; the thicker the haze, the whiter the sky. In the thickest hazes near deserts or industrial areas no blue may be visible. The blueness of the sky is thus a direct measure of the content of minute solid particles in the air.

Apart from their reddening near sunrise and sunset, both the sun and the moon may appear coloured on rare occasions, especially blue and green. The effect is produced by diffraction through diffuse clouds of dust or smoke, whose particles are of roughly constant size, such as may appear in the upper troposphere after violent volcanic eruptions or extensive forest fires. In detail, the phenomenon is not well understood.

8.4.2 Mirages

When light passes through layers of the atmosphere with different densities it may be refracted to give a *mirage*, of which there are two types:

(a) *Inferior mirage*, when objects near the horizon appear to be reflected in a pool of water to give an image just below the horizon. It may be seen over any extensive flat surface which has a temperature much greater than the air above it, usually as a result of insolation. Such mirages are popularly associated with deserts, but they can be seen in Britain quite easily over beaches, roads and airfield runways, for example. Light from an object on the horizon passes through the air in immediate contact with the ground at grazing incidence and, since this air is much warmer and less dense, it refracts the light upwards so that it can enter the eye giving an image below the original object. If the angle of incidence of the ray as it enters the lowest heated layer is too large, refraction is insufficient to prevent its intersecting the ground and so giving no image. Hence, only those objects just near the horizon can give a mirage. The effect of apparent reflection in a water surface is heightened if part of the blue sky light also produces a mirage. It is heightened further by shimmering brought about by shallow convection currents resulting from the very large lapse rates in the greatly heated layer a centimetre or so above the ground.

(b) *Superior mirage*, when objects near the horizon appear to be extended vertically above their true positions, or even objects beyond the horizon can be made visible and seem to 'float'. It may be seen over any extensive flat surface which is much colder than the air above it, for example, over land after a clear, calm night (especially snow-covered land) or over the sea when much warmer air flows across it (dry air, otherwise fog may form). A well-marked temperature inversion then exists near the surface. Light from distant objects is bent downwards near the top of the cold layer of air and on entering the eye gives an apparent image above the object. It is less common than the inferior mirage.

8.4.3 *Whiteout*

In polar regions when the ground is completely snow-covered and a uniform layer of cloud is present, say *St neb*, the intensity of diffuse, reflected light from the ground may be the same as that from the cloud, so making the horizon indistinguishable. The lack of shadows and structural detail blends the ground and sky into a whole to produce a *whiteout*.

BIBLIOGRAPHY

BALK, J.G.; 1952. Rare parhelion seen at Oxford, and a note on the frequency of solar and lunar halos and associated optical phenomena, 1882–1951. *Met Mag*, 81, pp. 263–266.

BIGG, E.K.; 1974. A glory in the stratosphere. *Weather*, 29, pp. 328–332.

GOLDIE, E.C.W.; 1971. A graphical guide to haloes. *Weather*, 26, pp. 391–393.

GOLDIE, E.C.W. and HEIGHES, J.M.; 1968. The Berkshire halo display of 11 May 1965. *Weather*, 23, pp. 61–69.

HATTINGA VERSCHURE, P.P.; 1973. Uncommon optical phenomena at Venlo 2 May 1972. *Weather*, 28, pp. 300–306.

LILJEQUIST, G.H.; 1956. Halo-phenomena and ice-crystals (Maudheim, 71°03'S, 10°56'W). Norwegian–British–Swedish Antarctic Expedition, 1949–52. Special Studies. Scientific Results, 2, Part 2, Oslo.

MALKUS, W.V.R.; 1955. Rainbows and cloudbows. *Weather*, 10, pp. 331–335.

SHUTE, C.C.D.; 1976. The 'Blue Moon' phenomenon. *Weather*, 31, pp. 292–296.

THOMPSON, A.H.; 1974. Water bows: white bows and red bows. *Weather*, 29, pp. 178–184.

PART II

SYNOPTIC METEOROLOGY

CHAPTER 9

AIR MASSES AND FRONTS

9.1 WEATHER CHARTS

9.1.1 *Introduction*

In Part I of this book we have considered many aspects of the weather from the point of view of a single observer—noting how such features as temperature, pressure, wind and clouds are constantly changing, and also something of the physical processes controlling these changes. These changes are of two types: firstly, there are those which are produced by alterations of the weather on the spot, for example, temperature changes resulting from insolation, or gustiness of the wind caused by turbulence; secondly, there are those which have been brought about by advection of different weather which has been formed elsewhere. We have, in fact, both 'home-produced' and 'imported' weather, and the weather experienced at any given place or time is a mixture of both. In technical language, these two types of changes are said to be brought about by 'development' and 'advection' respectively.

It follows, then, that the rate of change of a weather element observed at a fixed place—known as its *tendency*—comprises two parts, resulting from *development* and *advection*. In Part I we were primarily concerned with development; in Part II we shall consider advection and the results of blending the two together.

To understand how advection can alter the weather at any given place we need to know two things: the distribution of weather around that place, and the pattern of winds which moves the weather about. A convenient way of finding the distribution of weather is to have a large number of observers scattered evenly over the area, each making observations simultaneously and at regular intervals. Inspection of the observations then gives an overall picture of the weather; its clarity will depend upon many factors, in particular, on how closely the observers are spaced, how representative of the weather is each observation, and the skill of the observers themselves.

9.1.2 *Synoptic charts*

When we need to survey the weather over very large areas, even, say, the whole of the northern hemisphere, the number of necessary observations becomes very large and the only simple way by which they can be compared and understood is by plotting the information on a chart in some shorthand form. Some elements of the weather such as temperature, pressure and dew-point, which can be expressed numerically, are plotted in figures. For other elements, such as cloud and precipitation forms, symbols are used. Such a chart, for a fixed time and area, is known as a *synoptic chart*. It is an attempt at getting a 'bird's-eye view' of the weather, and is a most important tool, used both by weather forecasters and research meteorologists. Its main disadvantage lies in the method used to obtain the data for its construction. Since this is just a sampling technique it is clear that wherever the weather detail is small, as in a shower, much of it is easily missed because observers are usually spaced well apart, say 30 to 60 kilometres. Hence, only general patterns of the weather can be obtained. Locally, however, more detail can be filled in by the use of a meteorological radar (see Appendix A, Section A.2.2). Fortunately, most of the patterns in which we will be interested are larger than the spacing of the observers so that they can be seen clearly on synoptic charts.

We normally refer to two types of synoptic chart: 'surface' and 'upper-air' charts, referring respectively to those compiled from observations made by an observer on the ground, and those based on observations in the free atmosphere. In this book the term 'synoptic chart' will refer to a surface chart unless otherwise stated.

There are three stages in the preparation of a completed synoptic chart: *observation, plotting* and *analysis*. The first two cannot be considered in detail in this book. For further information the reader is referred to the *Observer's Handbook*, the *Handbook of meteorological instruments*, and *Instructions for the preparation of weather maps*.

After plotting the observations a synoptic chart is completed by a third stage—*analysis*, the comparison of neighbouring observations, and the identification of patterns among them. These patterns are then indicated by drawing either *isopleths*, lines joining places with equal numerical value of some given property (for example, isobars and isotherms), or symbols for those properties which cannot be expressed numerically. Analysis is a skilled and difficult operation. It belongs to the realm of the forecaster and researcher rather than the observer, so in this book we must be content to accept the results of analysis in the form of completed synoptic charts. Even these charts, which we will be using as illustrations, will be very greatly simplified. For those readers who wish to learn about the techniques of analysis, a useful introduction will be found in the book *The practice of weather forecasting*; other and more advanced references are given in the reading list.

9.2 THE GENERAL CIRCULATION

9.2.1 *The mean pressure distribution at sea level*

A synoptic chart gives a picture of the weather at a fixed moment in time. Another type of chart can be constructed showing the distribution of mean values of a particular feature of the weather, taken over a period of a year, say. Such charts showing, for example, mean temperature or mean duration of sunshine, are used extensively by climatologists. It is of interest to us to study one of these charts—that which shows the annual mean pressure over the northern hemisphere—because the distribution of mean pressure governs the pattern of mean winds in a way we can predict by using Buys Ballot's Law.

The chart of mean annual pressure at sea level shows four large features—belts of relatively high and low pressure extending approximately parallel to the equator and encircling the hemisphere. These belts, with the approximate latitudes within which they lie, are:

(a) *Equatorial low*, 0° to 10°N
(b) *Subtropical high*, 10° to 40°N
(c) *Temperate low*, 40° to 70°N
(d) *Polar high*, north of 70°N

Similar belts are found in the southern hemisphere. Careful examination of the chart shows that the belts are not continuous around the earth but each is split into a number of separate centres. Thus lows are prominent over the Iceland and Aleutian areas, and highs over the Azores and the central Pacific. In winter, the intense cooling of the atmosphere over land at high latitudes causes high pressure to develop over Canada and Siberia, interrupting the temperate low-pressure belt. In summer, the intense heating over subtropical latitudes causes depressions to form over the southern part of Asia and the south-western United States of America, so interrupting the subtropical high-pressure belt.

Even a synoptic chart shows many of these features, but the positions and intensities of the individual pressure systems vary markedly from day to day.

9.2.2 *The mean surface wind distribution*

The pressure patterns described in Section 9.2.1 are, of course, accompanied by corresponding wind patterns which can be divided into three zones:

(a) *Trade winds* (north-east trades in the northern hemisphere, and south-east trades in the southern hemisphere) between the equatorial low- and subtropical high-pressure belts. In some places the trades of one hemisphere cross the equator into the other hemisphere, especially when it is summer in the latter.

(b) *Temperate westerly winds*, between the subtropical high- and temperate low-pressure belts.

(c) *Polar easterly winds*, between the temperate low- and polar high-pressure belts.

These wind patterns are modified by such seasonal changes in the pressure distribution as mentioned above, so that, for example, the Siberian anticyclone in winter intensifies the north-east winds over southern Asia, and the low over southern Asia in summer reverses the normal winds to south-westerlies. These modifications, or *monsoons*, are really large-scale land- and sea-breezes, but they have an annual cycle, not a diurnal one as with the normal sea-breeze.

The term *general circulation* is applied to the wind systems of the world viewed as a whole. Figure 54 shows the general circulation of the surface winds over the northern hemisphere in a very diagrammatic way. In it, the effects of the presence of continents and oceans have been ignored. It will be seen that masses of air with contrasting temperatures are brought close together by these

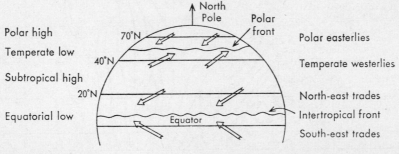

FIGURE 54. *Diagrammatic outline of the general circulation in the northern hemisphere*

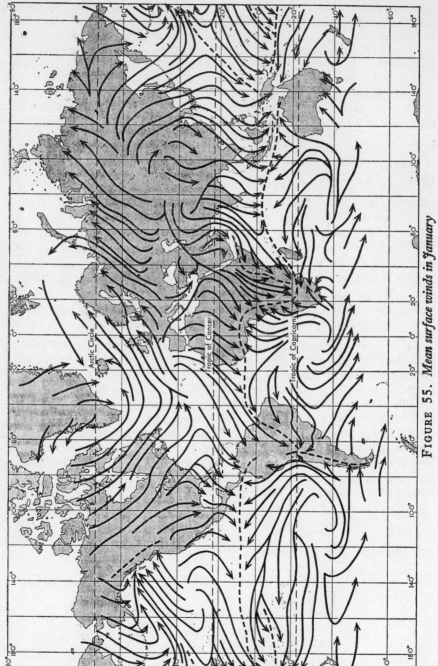

FIGURE 55. *Mean surface winds in January*

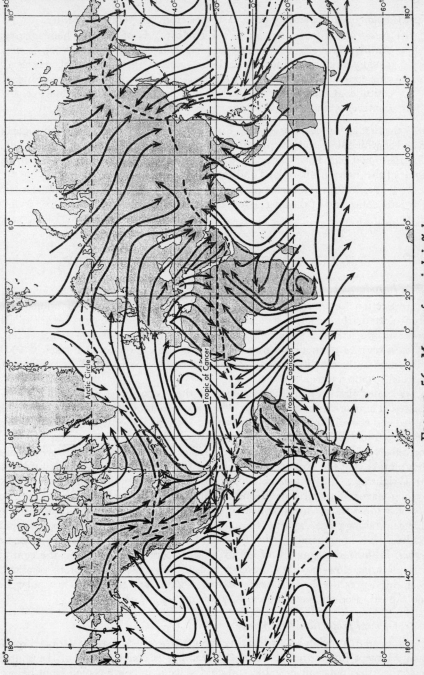

FIGURE 56. *Mean surface winds in July*

world-wide wind systems, and their boundaries are marked by frontal zones (see Section 1.6) which stretch around the world. There are two of these frontal zones:

(a) *Polar front*, separating polar easterlies from temperate westerlies.
(b) *Intertropical front*, separating the trade winds of the two hemispheres. An alternative term, the intertropical convergence zone (ITCZ), is now usually preferred since it has been realized that the characteristics of this zone are quite different from those associated with a front of higher latitudes. Over the oceans the range of movement is small, but over the continents it may be very large.

Only the first of these affects Britain. Figure 54 should be compared with Figures 55 and 56 which show the actual mean surface wind circulation in January and July. The more complex flow, particularly in the northern hemisphere, is mainly a result of the distribution of land and sea.

In the remainder of this chapter we shall be studying the weather characteristic of frontal zones and of the air masses which they separate.

9.3 AIR MASSES—A GENERAL SURVEY

9.3.1 *Introduction*

In Section 1.6 we saw that a chart on which temperature observations have been plotted from a large area usually shows several air masses separated by frontal zones, all on a scale of hundreds or even thousands of kilometres. An *air mass* may be defined as a mass of air with dimensions of the order of 1000 kilometres and with little or no horizontal variation of any of its properties, especially temperature.

We must first explain the existence of these enormous warm and cold air masses. Their origin becomes clear when we remember that the atmosphere is heated and cooled almost entirely by contact with the underlying surface. A warm air mass must be produced by prolonged contact with a warm surface and, conversely, a cold air mass with a cold surface. The heat transfer processes are slow: it takes days or even weeks to alter the temperature of an air mass by 10 or 20 °C right up to the tropopause. Over such a prolonged period, direct loss of heat by long-wave radiation can contribute significantly to the cooling of an air mass. So not only must the air mass be heated or cooled, but it must *stagnate* whilst the change is taking place. A part of the earth's surface where air masses stagnate, and gradually assume characteristics typical of that surface, is known as a *source region*. Prolonged stagnation of an air mass is commonly found in the central parts of large, slow-moving anticyclones, especially those in the subtropical and polar high-pressure belts. Hence, source regions are usually either polar or subtropical, producing corresponding *polar* or *tropical air masses*.

9.3.2 *General properties of air masses*

In its source region an air mass has properties which may be deduced quite easily. Except for diurnal changes, the stability is characteristic of the source region, so that polar air is stable and cold since it is being cooled from below, whereas tropical air is unstable and warm because it is being heated from below.

The anticyclones in the source regions are not permanent features; they undergo temporary changes which are associated with movement of the air masses to other parts of the earth's surface. When this occurs, an air mass is not in thermal equilibrium with the underlying ground, so that an air mass which is migrating from its source region is progressively modified. As examples we may note:

(a) An air mass having a *maritime* track will tend to become saturated especially in its lowest layers; one with a *continental* track will retain a dew-point similar to its original value because little water is available over land for evaporation.

(b) Polar air flowing to lower latitudes is warmed from below and becomes unstable; tropical air flowing to higher latitudes is cooled from below and becomes stable.

(c) An air mass influenced by divergence in its lowest layers, near a developing anticyclone for example, sinks slowly—a process known as *subsidence*—with the result that much of the troposphere becomes warmer, drier and more stable. On the other hand, an air mass influenced by convergence in its lowest layers, near a developing depression for example, rises slowly with the result that much of the troposphere becomes cooler, moister and less stable.

From these examples we see that the properties of an air mass depend upon:

(a) its *source region*,
(b) its *track* over the earth's surface,
(c) the extent of *divergence* and *convergence*.

These factors are useful as a basis for the classification of air masses; we shall use a simple system, based only upon source region and track, in which there are four basic air masses: *polar maritime (Pm)*, *polar continental (Pc)*, *tropical maritime (Tm)* and *tropical continental (Tc)*. Thus, to say, for example, that *polar maritime* air covers Britain means that the air mass has had a polar source and a maritime track before reaching Britain.

The idea of air masses was first introduced into meteorology by Bergeron in 1928. A thorough survey of British air masses was made by Belasco in 1952 (*Geophysical Memoirs* No. 87).

9.3.3 *The position of Britain*

The source regions and tracks of air masses approaching Britain are shown in Figure 57. Our position on the west coast of a continental land mass means that air masses flowing directly from the west are maritime, whereas those from the east are continental, except in so far as they may be modified by the Baltic, North and Mediterranean Seas. It is well known that our winds are predominantly from a westerly point so it is not surprising, therefore, that maritime air masses are the most common over Britain. Most source regions are several thousand kilometres away so that air masses take several days to reach us, during which time they become considerably modified. This is fortunate for us, because the contrast between polar and tropical air masses is lessened so that the weather changes accompanying an air-mass change are less abrupt. Places with a latitude similar to western Europe but on the *eastern* sides of large continents, for example,

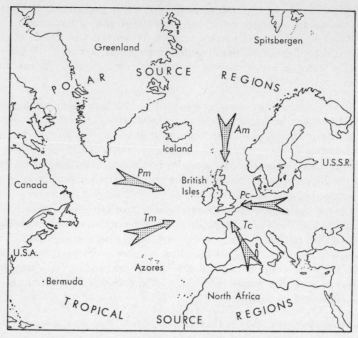

FIGURE 57. *Typical source regions and tracks of air masses found over Britain*

Am: arctic maritime; *Pm*: polar maritime; *Pc*: polar continental; *Tm*: tropical maritime; *Tc*: tropical continental.

in the New England states of the United States of America, and in Japan, experience far greater extremes because the air masses crossing them are relatively less modified.

The properties of an air mass depend upon the season to some extent. In the next two sections we shall study and compare British air masses in winter and in summer, confining our attention to those which come fairly directly from their source regions. It is important to realize that many air masses found over Britain have reached us by complex routes so that their properties may be intermediate between those to be described. Also, local and temporary source regions may be established over or near Britain, resulting in new air masses.

9.3.4 *Transformation of air masses*

When an air mass spreads to latitudes far from that of its source it may become so modified as to warrant reclassification. This applies particularly to the lowest few kilometres which, after all, are the first to be altered by contact with the new underlying surface. Thus, polar air flowing southwards, to below 30°N say, may become transformed into tropical air, and this is particularly easy because at such low latitudes the polar air is usually present only as a shallow layer near the ground, not extending up to the tropopause. The transformation of tropical into polar air takes place more slowly and through a deeper layer, largely by radiational processes.

9.4 AIR MASSES OVER BRITAIN IN WINTER

9.4.1 *Polar maritime air (Pm)*

Polar air with a source region in the northern parts of North America has a long ocean track before reaching Britain. Although originally very cold, say, -20 to $-40\,°C$ near the ground, the ocean, being relatively much warmer, produces great changes: the air mass is warmed throughout the troposphere, especially in the lower layers, and evaporation maintains a relative humidity of about 60 to 80 per cent. So, approaching our western shores, *Pm* air is cool, unstable, but not too moist.

Over the open sea and windward coasts convective clouds and showers are widespread, the showers being frequent and heavy at times, both by day and by night. Brief hail showers and thunderstorms may even occur. Such a storm is seen in Plate XI. *Pm* air streams in from the west usually between a low near Iceland and a high near the Azores as shown diagrammatically in Figure 58(*a*). That part of the airstream nearest the low usually has the largest and most developed clouds and showers; farther south subsidence near the high often produces an inversion based around 2 kilometres, so that convection is limited to below this level with the consequence that clouds are shallow (*Cu*, with *Sc* or *Ac cugen*) and any showers are light and scattered. If the airstream is flowing rapidly it

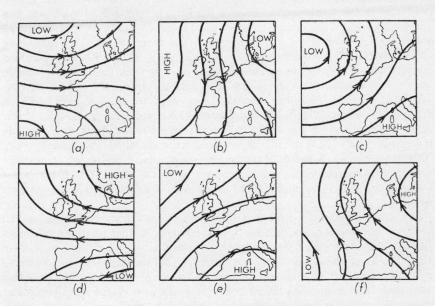

FIGURE 58. *Typical synoptic patterns associated with the commonest air masses over Britain*

Each map shows simplified isobars and directions of the geostrophic wind.
    (a) Polar maritime air (*Pm*) from the west.
    (b) Arctic maritime air (*Am*) from the north.
    (c) Returning polar maritime air (*rPm*) from the south-west.
    (d) Polar continental air (*Pc*) from the east.
    (e) Tropical maritime air (*Tm*) from the south-west.
    (f) Tropical continental air (*Tc*) from the south-east.

may reach Britain still sufficiently cold to allow some of the showers to fall as snow, especially over high ground, and this applies particularly to northern Britain.

These conditions apply to *Pm* air over the open ocean and windward coasts. Inland, night-time cooling is often sufficient to allow a stable layer to form near the ground which is destroyed by day-time heating, but only for a short period of a few hours around the time of maximum temperature. This stable layer prevents the formation of new convective clouds. Away from the western coasts, then, there is no cloud at night except where the decaying remains of any shower clouds drift across from the west—these would be *Sc* or *Ac cugen* and *Ci spi cbgen*. During the day, convective clouds are weak with few, if any, showers. Shelter by the high ground of western Britain also helps to keep clear skies at night. These clear skies promote maximum loss of heat by long-wave radiation, so night temperatures are low and air frost is common by dawn, perhaps with a few patches of radiation fog. Away from fog and showers visibility is good, especially in the west, because pollution by smoke is unlikely. It is clear from this outline that diurnal variations are important in *Pm* air over the land.

There are two varieties of *Pm* air which are particularly important. When *Pm* air flows from the north it has a short sea track from its source region around Spitsbergen or the Arctic Ocean and is therefore colder when it reaches Britain, so much colder (often near 0 °C) that the air mass is sometimes designated *Arctic maritime* (*Am*). It is very unstable, particularly if the low-pressure centre producing the northerly winds lies over southern Scandinavia or the North Sea (Figure 58(*b*)), and the resulting showers are frequent and heavy and usually of snow. They are best developed over our north and east coasts where the strong winds can cause drifting of lying snow. Some of the most severe snow conditions in these areas are found in an *Am* air mass flowing from the north, especially over high ground where orographic ascent of the air adds to the shower activity. Weaker showers are also found on exposed west coasts, for example, over Northern Ireland, north and west Wales, and Cornwall, but well inland showers are few and far between. Very low temperatures, below −10 °C and even occasionally below −20 °C, can be found at night in sheltered spots which have a snow cover. The long clear nights and poor conducting properties of fresh lying snow cause rapid falls in the temperature of the topmost layer of the snow. Our first and last snows of the winter over high ground in the north are often produced by this air mass.

A slow-moving depression west of Britain forces *Pm* air into latitudes well south of 50°N, sometimes even south of the Azores, so that when it approaches us from the south-west it has been greatly modified—it is warmer and moister than the *Pm* air which has come to us directly. This air mass, which is returning to higher latitudes as it crosses Britain, is known as *returning Polar maritime* air (*rPm*). As it approaches Britain from the south-west (Figure 58(*c*)) it crosses progressively cooler water so that by the time it reaches our shores its lowest layers have been cooled again and are stable. This cooling prevents the formation of convective clouds and, together with the accompanying moistening, may be sufficient to produce a thin layer of cloud of the *Sc str* type at about 500 metres. This cloud spreads inland and often is only partially broken down by the effects of day-time heating or shelter by high ground. Only the surface layers are stabilized; aloft, the air mass is still unstable so that *Cb* and showers may develop as the air mass flows over high ground, especially in the west. (See also Sections

10.2.3 and 10.3.2 for *Pm* air in cold pools and polar lows; and Sections 11.2 and 11.3 for *Pm* air in anticyclones.)

9.4.2 *Polar continental air (Pc)*

Polar air with a source region in Siberia or eastern Europe reaches us when easterly winds blow across Europe under the influence of an anticyclone over Scandinavia (Figure 58(*d*)). Although very cold (maximum temperatures can be around 0 °C), subsidence within the anticyclone warms and dries the upper parts of the troposphere so that an inversion forms with its base between 1 and 3 kilometres. Below this the air is warmed and moistened as it crosses the North Sea, so that although it may be cloud-free over the continent, there are usually large amounts of *Cu* and shallow *Cb* over our east coasts, with showers of rain or snow. Snowfall can be considerable over the immediate coastal regions, and Kent is likely to suffer particularly badly because of the lower temperatures there and the greater depth of the unstable layer (farthest from the high-pressure centre). Diurnal variation of shower activity is small near the coast, but inland, showers are largely day-time features and usually few anyway, especially to the lee of high ground, for example, in south-west England, west Wales and west Scotland.

A *Pc* airstream persisting for several days gives us our most severe winter weather—especially with regard to low day-time temperatures which are most noticeable when accompanied by strong easterly winds.

9.4.3 *Tropical maritime air (Tm)*

Tropical air with a source region over the North Atlantic in the latitudes of the Azores and Bermuda reaches us from the south-west, often under the influence of an anticyclone near the Azores or western Europe (Figure 58(*e*)). It is warm, and very moist in its lowest layers, but above about 1 kilometre it is often subsided, warm and dry, so that clouds occur mostly in the lowest kilometre. As it spreads towards Britain *Tm* air is cooled from below by contact with progressively colder sea surfaces so that it becomes stable and saturated, or nearly so, in the lowest kilometre. This layer is often filled with a very extensive sheet of *St*, or even fog—advection fog; local outbreaks of drizzle are common too. Over high ground there is much hill fog and the drizzle is often more pronounced. Inland the cloud is sufficiently thick to prevent the weak winter insolation from dispersing it and all that may be observed is a temporary day-time rise in cloud base. However, descent in the lee of high ground, for example, over the west Midlands, Yorkshire and around Aberdeen and the Moray Firth, may cause breaks to appear.

This mild and damp air mass gives us our typical muggy days in winter, when the temperature may reach 10 to 12 °C even in January, and on rare occasions 15 °C. Visibility is poor, especially in drizzle and near industrial areas. Diurnal variations of weather in *Tm* air are small in winter. (See also Section 10.1 for *Tm* air in depressions.)

9.4.4 *Tropical continental air (Tc)*

Tropical air from a source region over North Africa does not affect Britain in winter, at least not near the ground, for two reasons: firstly, the necessarily prolonged south to south-east winds are uncommon; and secondly, the air mass is so modified on crossing the warm Mediterranean Sea, and then by the cold continent of Europe, that it merely resembles a mild form of *Pc* air on reaching Britain.

9.5 AIR MASSES OVER BRITAIN IN SUMMER

9.5.1 *Polar maritime air (Pm)*

Over the open sea and windward coasts summer *Pm* air resembles the winter form, but the instability is less and convection weaker; if is of course also warmer, with temperatures 12 to 15 °C where unaffected by diurnal variations. Inland the properties of *Pm* air are markedly different in summer compared with those in winter. Intense insolation causes widespread convection by day with heavy showers of rain and hail and with thunderstorms, especially in the south-east half of England. Typical clouds are illustrated in Plates VIII and X. In summer *Am* air gives us noticeably cool spells, often with frequent and heavy showers, whereas *rPm* is often as warm and moist as *Tm* air near the ground, but is, of course, still unstable aloft. Inland, day-time heating is often capable of dispersing the layer cloud which has spread from the south-west, replacing it by towering *Cu* and *Cb* with squally showers and thunderstorms. Some of our most severe thunderstorms occur in moist, unstable *rPm* air masses.

9.5.2 *Polar continental air (Pc)*

Polar air from the east or south-east in summer flows over the warm continent and reaches us as a warm, dry air mass, often cloud-free. Cooling and moistening by the North Sea may give advection fog or *St* cloud over our east coasts. Diurnal temperature changes are often marked.

Pc air resembles *Tc* air and, indeed, may be considered as only a rather cooler form of that air mass.

9.5.3 *Tropical maritime air (Tm)*

The properties of *Tm* air in summer over the open sea and windward coasts are similar to those of winter, except that it is rather warmer, 15 to 18 °C compared

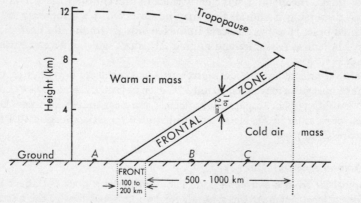

FIGURE 59. *Vertical cross-section through a frontal zone*

with 8 to 10 °C. Inland, day-time heating is often sufficient to disperse the *St*, leaving either clear skies or small amounts of *Cu hum*. Only rarely is it unstable enough at higher levels to allow large convective clouds to form. Temperatures can reach 20 to 30 °C by day; at night radiation fog may form but soon disperses after dawn by insolation.

9.5.4 *Tropical continental air* (*Tc*)

Air resembling true *Tc* from the Sahara sometimes forms over south-east Europe in summer, and can reach us under the influence of an anticyclone over north-east Europe (Figure 58(*f*)). It is hot and unstable, but its dryness prevents formation of much convective cloud. Our so-called 'heat-waves' occur with this air mass, when day maxima may reach 25 to 35 °C; night minima are correspondingly high, 15 to 20 °C. *Tc* air is often hazy as a result of both dust and smoke from industrial areas on the continent.

9.6 FRONTS—A GENERAL SURVEY

9.6.1 *Some definitions*

A *front* is the intersection of a frontal zone with the ground, and a *frontal zone* is the boundary between two adjacent air masses, extending into three dimensions. The particular property we use to distinguish the air masses will also indicate the front, and since temperature is commonly used for this purpose we shall use it to define a frontal zone. It is a zone with a large horizontal temperature gradient—large, that is, compared with the gradients within the air masses. It must be remembered, however, that other properties may be used to distinguish air masses, for example, dew-point, and we can therefore speak of such features as 'dew-point fronts'. In this book we shall consider the terms 'front' and 'frontal zone' as referring to boundaries marked by temperature gradients.

A frontal zone, as shown in Figure 59, slopes upwards above the colder air mass which thus lies in the form of a thin wedge below the warmer air. The slope of this wedge is small, about 1 in 100, so a frontal zone which extends upwards through the whole troposphere will meet the tropopause along a line about 1000 kilometres on the cold side of the surface front. The vertical depth of the frontal zone, which is caused by a slow mixing of the adjacent air masses, is usually 1 to 2 kilometres, so its horizontal width is 100 to 200 kilometres. If a radiosonde ascent were made at a place represented by *B* in Figure 59, the two air masses would be indicated quite clearly by the vertical distribution of temperature, which might be as shown in Figure 60. We can see the cold air mass below surmounted by the warm air mass, and between them over a depth of 1 to 2 kilometres is the frontal zone—here it is shown as an inversion. In less extreme examples it would be represented by a lapse rate which, although still positive, is smaller than the lapse rates found within the air masses above and below.

A frontal zone is said to be sharp if the temperature gradient is well defined; on some occasions it may be so sharp that the temperature is almost discontinuous between the two air masses. The frontal zone is diffuse if the gradient is ill defined.

The formation of a front, or the sharpening of one already existing, is known as *frontogenesis*; it occurs whenever convergence exists in a region near the ground where the air already has a temperature gradient, even though this gradient

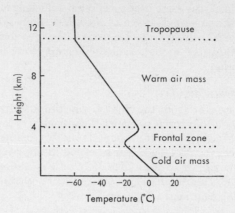

FIGURE 60. *Vertical distribution of temperature through a marked frontal zone as may be measured by a radiosonde ascent at B in Figure 59*

may be initially quite small. In fact, convergence is essential in maintaining a sharp front, so it is not surprising that fronts are generally most marked near developing depressions. The weakening or destruction of an existing front is known as *frontolysis*; it occurs when the front is subjected to divergence. Hence fronts are generally weakest near developing anticyclones.

Air masses are generally in motion, and since the wind cannot blow through a front, it too will be in motion. A front moving in such a way that warm air replaces cold air when it passes a given place is said to be a *warm front*; if cold air replaces warm air it is a *cold front*; if the movement is small and irregular the front is said to be *quasi-stationary*. These widely used terms were introduced in 1922 by J. Bjerknes and Solberg. On synoptic charts the different types of fronts are represented by conventional symbols as shown in Figure 61.

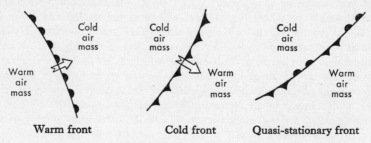

Warm front   Cold front   Quasi-stationary front

FIGURE 61. *Conventional symbols used to represent fronts*

9.6.2 *Horizontal winds near fronts*

On synoptic charts it is found that where isobars cross a front they usually do so with a change in direction, and this change is such that the acute angle produced always points towards the direction of higher pressure (see Figure 62). This is true whether the front is of warm or cold type. The greater the contrast in temperature between the air masses, the more acute is the angle, so that a

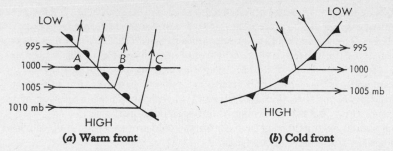

(*a*) Warm front   (*b*) Cold front

FIGURE 62. *Diagram illustrating the bending of isobars as they cross a front*

weak front causes little bending of the isobars crossing it whereas a pronounced front is indicated by a well-marked change in orientation of the isobars.

Since the isobars lie in different directions on each side of the front, it follows that the winds must blow in different directions correspondingly. In Figure 62 the geostrophic wind directions are indicated by the arrows on the isobars. It can be seen that as a front moves over an observer, the wind veers; always there is a veer, whether the front is warm or cold.

We can find an explanation of the observed differences on each side of a front in the following way. Consider three radiosonde stations, *A*, *B* and *C*, lying along the line shown on the surface chart in Figure 62(*a*). These stations are also indicated on the cross-section through the frontal surface shown in Figure 59. An ascent at *A* will be wholly in the warm air mass and would give a temperature distribution as shown by curve *A* in Figure 63. A similar result would be obtained from an ascent at any other point along the line as long as it lies within the warm air, that is, west of the front. Ascents at *B* and *C* will be partly in the cold air (at lower levels) and partly in the warm air (aloft), as shown by curves *B* and *C* in Figure 63, where the greater depth of cold air at *C* is clear. Now we have seen in Section 2.1.2 that the atmospheric pressure at the ground increases as the mean temperature of the atmosphere decreases, so that, since the important difference between curves *A* and *B* is that the lower levels of *B* are colder than *A*,

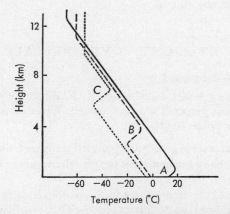

FIGURE 63. *Diagram showing the vertical distribution of temperature at points A, B and C in Figures 59 and 62(a)*

it follows that the pressure at the ground at B is greater than at A. Also, since the depth of cold air at C is greater than at B, the pressure at C must be greater than at B. From this, we see that the atmospheric pressure at the ground is constant at points along the line to the west of the front but to the east it increases progressively towards C as the depth of cold air increases. Isobars can then be drawn only as in Figure 62(a). This argument is strictly true only if pressure aloft in the warm air is constant in the direction AB. Since this is seldom so, the variation of surface pressure along AB can be explained only partially in this way.

A similar argument may be used for the cold front in Figure 62(b) where the cold air lies to the west of the front.

To some extent this pressure distribution is counteracted by a gradual lowering of the tropopause from the warm air mass into the cold. The accompanying warmer stratosphere tends to give a lower pressure at the ground in the cold air mass.

9.6.3 Vertical winds near fronts

A frontal zone is usually associated with *vertical* motion of the air masses in its vicinity. As was seen in Section 9.6.1, convergence near the ground is essential in the formation and maintenance of a sharp front. Now convergence near the ground is associated with upward motion (see Section 2.6.3) leading to the formation of clouds as a result of adiabatic expanison and cooling, provided that the air is moist enough. Thus a sharp or sharpening front is an active front—active in the sense that it will be associated with thickening cloud and, eventually, precipitation. The higher humidities normally found in the warm air, in Britain anyway, ensure that most of these clouds form in the warm air. The direction of the warm air's vertical motion enables us to distinguish two types of front: those where the warm air rises relative to the frontal zone, called *ana* fronts, and those where the warm air sinks, called *kata* fronts. Bergeron introduced these terms in 1937.

Most fronts are complex in that they may be ana or kata fronts at different levels or at different places along the fronts; their character may also vary with time. These complexities lead to a large number of possible cloud combinations; in fact, it is true to say that no two fronts are exactly alike.

We shall now consider the weather associated with model fronts of types which can be found over Britain.

9.7 FRONTS OVER BRITAIN

9.7.1 An ana warm front model

Figure 64 shows a model distribution of clouds and precipitation at an ana warm front in a plane perpendicular to the front. When using this diagram there are some important points to note:

(a) The vertical scale is greatly exaggerated compared with the horizontal.
(b) The arrows show the motion of the air relative to the front, which is itself moving.
(c) The whole of the troposphere in the warm air mass is shown as rising, but with varying rates at different levels. This, together with the variability of the relative humidity at different levels, explains the multi-layered nature of the frontal clouds at their upwind edges. Farther towards the

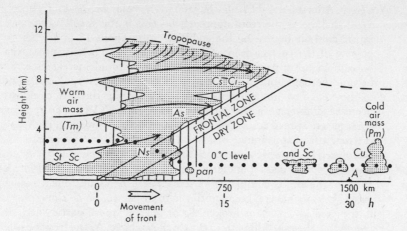

FIGURE 64. *Vertical cross-section through an ana warm front model*

The abscissa scales are horizontal distance from surface front and time before passing of surface front assuming constant speed of 50 km h⁻¹.

frontal zone, the cloud layers thicken and sometimes there are no cloud-free layers throughout most of the troposphere.

(d) The arrows which indicate motion of the air masses end at the frontal zone. This does not mean that the air comes to rest there, but that the component of the motion perpendicular to the front decreases as the air approaches the front, finally becoming zero. An accumulation of air behind the front might be expected as a result of this, but it does not occur because the wind component parallel to the front increases as the front is approached. In fact, this latter component is the larger at most levels, especially in the upper half of the troposphere where it can be as much as ten times the former. It follows that winds at these high levels blow approximately parallel to the front (see Section 10.1.4).

(e) Some of the cold air takes part in the vertical motion, but it is usually so dry from previous subsidence that not much cloud can form within it.

(f) Away from the front the weather is characteristic of the air masses; here, they are taken as *Tm* and *Pm*.

(g) The *Tm* air is shown as being stable, which is its usual condition, so the clouds in it are of a stratiform type. If the warm air were unstable, cumuliform clouds would be present.

(h) The *Pm* air lying immediately below the frontal zone in the middle troposphere is often remarkably dry with relative humidities below 10 per cent—this is the so-called *dry zone*.

The weather sequence experienced by an observer when such a front passes over him can be deduced by moving the diagram to the right and assuming no change in the existing weather distribution. If we take a frontal speed of 50 km h⁻¹ (say, 25 kn), which is a reasonable value, and also ignore diurnal variations, the weather changes observed at a place originally represented by point *A* in Figure 64 would be as follows:

(a) *Clouds*. At first *Cu* and *Cb* may be seen in the polar air, gradually becoming more shallow as the front approaches—this is the result of subsidence in the cold air increasing the stability aloft. About 15 hours before the passing of the surface front, *Ci fib* or *Ci unc* would be visible near the horizon in the direction of the front and it would increase steadily to become perhaps four oktas (four-eighths of sky covered) of *Ci* and *Cs* in the next few hours, whilst at the same time the low cloud would decrease. This stage is illustrated by Plates VI and XVII. The upper cloud continues to spread across the sky, thickening and lowering, perhaps with the temporary appearance of optical phenomena, but eventually the *Cs* is replaced by *As* which is so thick that the sun or moon is only visible as if through ground glass (popularly known as a 'watery sky'—see Plate VII). In turn, the sun is completely obscured and the cloud remains as either *As op* or *Ns* for several hours until the front passes. Some cloud may form in the cold air, perhaps broken *Sc* or *Ac*, but more particularly *St fra* (*pannus*) in the rain falling from the thick *Ns* above. As the front passes, the main cloud mass moves away and is replaced by the *St* or *Sc* typical of a *Tm* air mass.

(b) *Precipitation*. Showers may fall from the *Cu* and *Cb* in the cold air. The *Ci* and *Cs* give delicate trails of ice crystals which fall as *virga* but only for a short distance because they sublime in the drier, cold air mass. As the frontal cloud thickens, the precipitation becomes more intense and falls to lower levels, usually first reaching the ground some 4 to 6 hours before the passing of the surface front and then persisting until the front passes. Most of this precipitation starts as snow in medium and high clouds, having grown by the Bergeron process and by aggregation, but on falling to lower levels it may continue to grow by colliding with water droplets. Behind the front, any drizzle in the warm air forms in wholly liquid clouds by coalescence. The change in type of precipitation is clearly shown by a decrease in drop sizes but an increase in drop concentrations. The total fall usually amounts to 5 to 10 millimetres. If the polar air is cold enough snow may be able to reach the ground; this is most likely in the leading parts of the precipitation.

(c) *Temperature*. The change from polar to tropical air causes a temperature rise, usually about 5 °C, concentrated over a period of 1 or 2 hours whilst the frontal zone passes. Temperature may fall a little in the precipitation either because of evaporation or the melting of snow as it falls; prolonged snow can reduce the air temperature to very near 0 °C. Occasionally, snow which at first melts to rain whilst falling may subsequently be able to reach the ground unmelted as a result of this cooling.

(d) *Humidity*. The change from polar to tropical air causes a humidity rise, usually from 60 to 80 per cent to 90 to 95 per cent. Again this is concentrated mostly in the frontal zone. Evaporation of precipitation, however, causes the relative humidity to start rising well ahead of the front.

(e) *Visibility*. This is good in the polar air but only moderate in the tropical air; it is poor or bad in precipitation, especially in snow.

(f) *Pressure*. As the front approaches, pressure falls, slowly at first and then more quickly, perhaps at the rate of 1 to 2 millibars per hour. After the front has passed, pressure ceases to fall or it falls more slowly; sometimes it may even rise slowly. A total fall of 10 to 20 millibars would not be unusual. Not all of this is the result of simple replacement of the cold, dense air mass by the relatively warm and less dense air mass. On average, about half of the fall results from development of the depression with which the front is usually associated (see Sections 10.1 and 10.5).

(g) *Wind.* Most warm fronts over Britain move up from the south-west. Winds ahead of such a front are from the south or south-east; behind the front they veer to south-west or west. No matter from which direction the front approaches the observer, the wind *veers* as it passes, the greater part of the veer taking place in the one to two hours that it takes the frontal zone to pass over. Wind speed usually increases steadily as the front approaches, but may decrease again behind it.

From this outline of frontal weather we can see that the passing of an ana warm front is accompanied by fundamental changes, resulting not only from the change of air mass but also from the front itself. Most warm fronts over Britain are of the ana type, but they are often poorly developed (see also Section 10.1).

9.7.2 *A kata warm front model*

The fundamental difference between the kata warm front and the ana warm front is its generally downward-moving warm air, especially in the middle troposphere. The extensive medium- and high-level cloud sheets of the ana warm front are absent. Instead, as can be seen in Figure 65, much of the cloud is of the thick *Sc op* type, often wholly warmer than 0 °C, especially in summer, so that precipitation is light and in the form of fine rain or drizzle, both formed by coalescence. Both air masses usually contain much *Sc*, whose tops are limited by inversions formed at the base of the descending (subsiding) air in the middle troposphere. This *Sc* in the cold air is often formed by the spreading of *Cu* tops, accumulating in large amounts so that it is difficult to distinguish from *Sc op*; towards the front, indeed, it may be thickened into *Sc op* by convergence. In the warm air the *Sc* is usually thin and of the type *Sc str pe*.

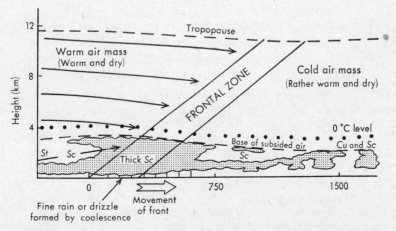

FIGURE 65. *Vertical cross-section through a kata warm front model*
The abscissa scale is horizontal distance from surface front (km).

Changes in other weather elements, such as temperature and humidity, are similar to those for an ana warm front but are smaller in amplitude and also spread over a longer time interval.

9.7.3 *An ana cold front model*

The distribution of clouds and precipitation with an ana cold front is simply the reverse of that found with an ana warm front as can be seen from Figure 66. But there is an important difference: the greater slope of the ana cold front— about 1 in 75 compared with 1 in 150 for an ana warm front—means that the weather is confined to a belt about half of the width of the belt found with the warm front. The rear edge of the cloud mass, found at high levels, is often very clear-cut and extends as an almost straight line across the sky. Weather changes accompanying an ana cold front are much more abrupt than for the warm front; occasionally, when the frontal zone is only a few miles across, the different winds in the two air masses may be inferred by observing the motion of low cloud on opposite horizons. Ana cold fronts are not common in Britain, and those which do occur are largely confined to the winter half of the year.

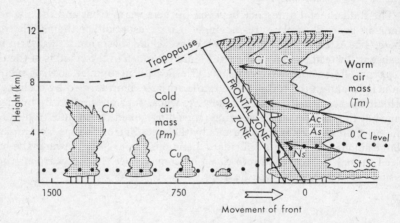

FIGURE 66. *Vertical cross-section through an ana cold front model*
The abscissa scale is horizontal distance from surface front (km).

9.7.4 *A kata cold front model*

The 'classical' cold front is normally shown as in Figure 66, but with one important difference—the stratiform cloud is replaced by *Cu* and *Cb*. Many cold fronts which cross Britain do not conform to this 'classical type'. There are two reasons for this: firstly, the warm air mass is seldom unstable and therefore unlikely to give cumuliform clouds (see Section 10.2.4); secondly, much of the warm air is found to be descending, giving predominantly cloud-free air in the middle troposphere (Figure 67). Weather changes accompanying the passing of such a kata cold front are similar to those observed with a kata warm front except that the order is reversed. Usually there is only a gradual replacement of the *St* and *Sc* of the warm air mass by more cumuliform cloud, and any precipitation is light, localized and of relatively short duration. Changes in temperature, humidity, visibility, pressure and wind are slow and ill defined.

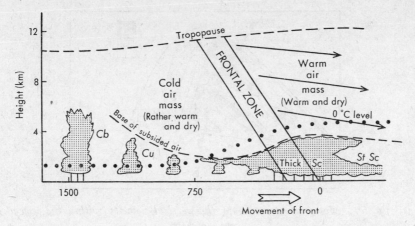

FIGURE 67. *Vertical cross-section through a kata cold front model*
The abscissa scale is horizontal distance from surface front (km).

9.7.5 *Some further ideas on fronts*

The concept of air masses and fronts has proved very useful in the analysis of synoptic charts and the models of the associated cloud and precipitation systems are helpful to forecasters provided the limitations of the models are realized. It must be remembered that each front is different from any other and there are no really 'typical' weather patterns associated with them. Fronts and their associated weather are to be regarded as products of the three-dimensional wind flow found in lows and highs.

Recently Doppler radar (see Appendix A, Section A.2.1) has been used to make detailed studies of air motions in precipitating systems. Probably the most significant concept to emerge from this work is that of the 'conveyor belt' illustrated in Figure 68. The conveyor belt is a well-defined flow of air, from about 100 to 1000

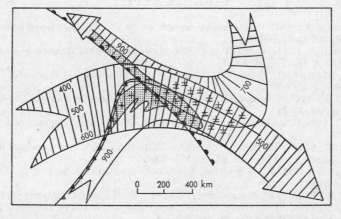

FIGURE 68. *Features of large-scale flow which determine distribution of precipitation at the surface*

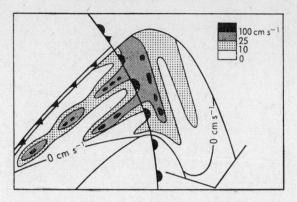

FIGURE 69. *Mesoscale structure of the vertical velocity within the ascending conveyor belt*

kilometres wide and about a kilometre deep which starts off at low levels parallel to and in advance of the cold front and gradually ascends as it approaches the surface warm front, turning to the right as it flows above the frontal surface to become nearly parallel to the front. Figure 68 shows the conveyor belt (stippled) ascending from 900 to 500 millibars (say 1 to 5 or 6 kilometres), the low-level flow ahead of the fronts descending from 700 to 900 millibars (3 to 1 kilometres) and the mid-tropospheric flow at about 500 millibars (5–6 kilometres). If the conveyor belt ascends above the cold-frontal surface the front is an ana front, otherwise it is a kata front. The rate of ascent is of the order of 10 cm s⁻¹ but with mesoscale variability as indicated in Figure 69. The regions of maximum vertical velocity associated with maximum rates of rainfall are of the order of 20 by 30 kilometres and are aligned parallel to the cold front in the warm sector and parallel to the warm front ahead of it. It seems likely that the conveyor belt and the frontal structure mutually interact.

BIBLIOGRAPHY

BROWNING, K.A.; 1971. Radar measurements of air motion near fronts. *Weather*, 26, pp. 293–304 and 320–340.

BROWNING, K.A. and HARROLD, T.W.; 1970. Air motion and precipitation growth at a cold front. *Q J R Met Soc*, 96, pp. 369–389.

FREEMAN, M.H.; 1961. Fronts investigated by the Meteorological Research Flight. *Met Mag*, 90, pp. 189–203.

GRANT, D.R.; 1967. Temperature and humidity fluctuations in a dry frontal zone. *Met Mag*, 96, pp. 33–41.

GRANT, K.; 1975. The warming and moistening of cold air masses by the sea. *Met Mag*, 104, pp. 1–9.

MILES, M.K.; 1962. Wind, temperature and humidity distributions at some cold fronts over SE. England. *Q J R Met Soc*, 88, pp. 286–300.

SANSOM, H.W.; 1951. A study of cold fronts over the British Isles. *Q J R Met Soc*, 77, pp. 96–120.

SAWYER, J.S.; 1955. The free atmosphere in the vicinity of fronts. *Geophys Mem*, 12, No. 96.

SAWYER, J.S.; 1958. Temperature, humidity and cloud near fronts in the middle and upper troposphere. *Q J R Met Soc*, 84, pp. 375–388.

WALLINGTON, C.E.; 1963. Mesoscale patterns of frontal rainfall and cloud. *Weather*, 18, pp. 171–181.

CHAPTER 10

DEPRESSIONS

10.1 WARM-SECTOR DEPRESSIONS

10.1.1 *Introduction*

Synoptic charts were first prepared during the middle of the nineteenth century when, for the first time, observations made simultaneously over a wide area could be collected quickly by using telegrams which at that time had been recently introduced. Since then, an enormous number of charts has been prepared, and over the years the area covered by them has steadily increased as more countries have participated in the making of observations, so that today it is possible to construct a synoptic chart covering the whole of the northern hemisphere. There are few areas where observations are not made; even the oceans, deserts and polar regions contribute to the continuous stream of information. Nowadays, radio and teleprinters have largely replaced telegrams as a means of reporting observations.

Simplified forms of synoptic charts are published daily in many countries for use by the general public. They usually take the form of daily weather reports; British *Daily Weather Reports* are obtainable from the Meteorological Office. Even a quick examination of a sequence of such charts covering, say, a period of a few months, shows that the patterns of isobars, winds and weather are constantly changing. The variations possible are endless; no two charts look alike—exact repetition does not occur. Although the changes are exceedingly complex it is possible to find definite life-cycles of the main pressure features— the lows and highs—each of which usually lasts a few days. During such a period, although it is short, a low for example may form, deepen rapidly, then fill and finally disappear. These developments are associated with rapid changes of the weather, and since the low may move 1000 or 2000 kilometres, the weather over large areas will be affected by its existence. Clearly, it will be of use to us to study some of the properties of these features of the pressure patterns. In this chapter we shall confine our attention to depressions and troughs, leaving anticyclones and ridges to Chapter 11. Our account will be largely descriptive for three good reasons: firstly, most of the studies which have been made in the past have been of this nature; secondly, most explanations of the observed properties of the highs and lows have been tentative and sketchy, and it is only in recent years that any attempts at rigorous explanations have been made; thirdly, a practical point, the structure of the weather patterns associated with pressure systems can be seen directly by careful observation.

Among the first serious descriptions of depressions, containing also some ideas of the physical processes involved, were those of FitzRoy in 1863 and Shaw in 1911. Although even in those early days some principles were introduced which have since been accepted as part of current theory, at the time they met with little favour. It was not until 1918 that the principles forming the foundation of modern ideas were introduced. In that year, J. Bjerknes described the way

153

by which most depressions of temperate latitudes developed on the polar front. Frontal depressions are common over and near Britain, but other non-frontal depressions occur and must also be considered.

There is a common misconception which should be mentioned here. A depression is not to be considered as just a cyclonic whirl of air, isolated from its surroundings. The air which takes part in its wind circulation passes *through* the depression, entering in one region and leaving at another. This is brought about in part by the upward motion of the air, normally found in a developing depression, such that air is drawn inwards at low levels and later ejected at high levels, and in part also by the movement of the pressure pattern, which movement may differ radically from the motion of its constituent airstreams. Only on rare occasions may a depression be even approximately like a self-contained whirl of air.

The characteristic properties of a depression—its centre of low pressure around which lie roughly concentric isobars, and is anticlockwise circulation of winds (in the northern hemisphere)—have been mentioned in Section 2.2.3. These are, however, characteristics which we have observed only on a surface chart. It must be realized that depressions are three-dimensional features, often extending upwards throughout the troposphere and lower stratosphere, and having structures which usually vary with height. Some idea of the wind patterns aloft corresponding to the changing pressure patterns with height may be observed directly from the movement of medium- and high-level clouds.

10.1.2 *Life-cycle of a model warm-sector depression*

In 1918 the results of intensive work by a group of Norwegian meteorologists were described by J. Bjerknes. These Norwegians found that the development of many middle-latitude depressions took place along the front separating polar from tropical air—the polar front. Since then the original ideas have been developed and modified, especially by J. Bjerknes and Solberg in 1922 and by Bergeron in 1928. Nowadays these ideas form the basis for weather analysis in temperate regions.

Consider the polar front over the North Atlantic lying, say, in an approximately east to west direction, separating *Pm* air to the north from *Tm* air to the south, and suppose both airstreams flow towards the east (Figure 70(*a*)). The front is quasi-stationary and may remain so for a day or two. Often, however, it is locally distorted by the formation of a bulge of warm air extending a little towards the cold air (Figure 70(*b*)) and moving along the front with a velocity approximately the same as that of the warm airstream. Such a bulge is known as a *frontal wave*.

The surface pressures at places near the wave are a few millibars below the corresponding values in the same places on Figure 70(*a*); hence, the isobars to the north of the wave bend southwards locally. Some may even intersect the front and so produce wind components across it sufficient to make the wave move eastwards as described. The leading edge of the warm bulge is a *warm front* and has been shown as such in Figure 70(*b*); conversely, the rear edge is a *cold front*. A frontal wave may move at about 15 to 20 metres per second (30 to 40 knots) and if it retains its shape without significant development it may travel 1000 kilometres per day, simply as a temporary disturbance running along the polar front. When this occurs it is said to be a *stable wave*; its passing has little permanent effect on the position of the polar front.

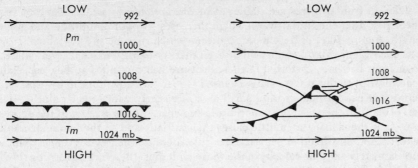

(a) Quasi-stationary part of the polar front.

(b) Frontal wave.

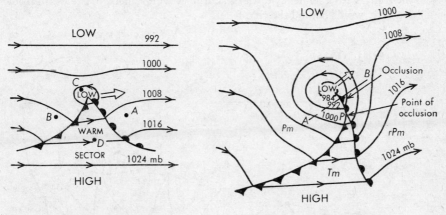

(c) Warm-sector depression.

(d) Partly occluded depression.

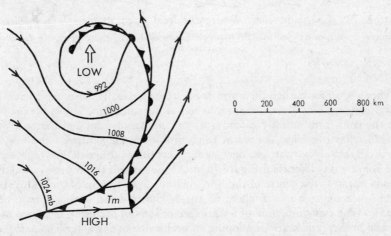

(e) Decaying depression.

FIGURE 70. *Life-cycle of a model occluding depression*
Arrows on the isobars show the direction of the geostrophic wind.

Not all frontal waves are stable; some are *unstable*, by which is meant they develop so that the initial warm bulge increases in both amplitude and wavelength. Figure 70(*c*) represents a condition which may be present about 1 day later than Figure 70(*b*), and it will be seen that the warm bulge is larger, and the polar front has been distorted farther. Pressure has continued to fall, especially near the tip of the wave where a closed isobar has now appeared—that is, a depression has formed—with a small area of easterly winds near point *C*. The area of warm air between the two fronts is known as the *warm sector* of the depression. In it the isobars are a little closer than initially and so the warm air moves faster; also their direction has backed a little. Because the low continues to move approximately with the velocity of the warm air it must therefore be turning slowly to the left of its original course. This is shown in Figure 71. Notice in Figure 70(*c*) how the isobars in the cold air mass have become progressively more distorted near the front, with a tendency towards southerly winds ahead of the warm front and northerly winds behind the cold front.

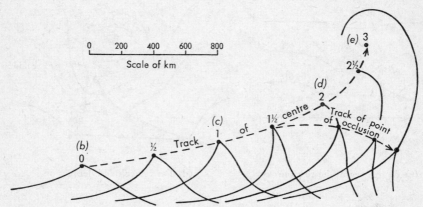

FIGURE 71. *Continuity chart showing the development of an occluding depression*
Letters correspond to sections (b) to (e) of Figure 70. Figures give the age of the depression in days.

Once a low has reached the warm-sector stage it usually continues to develop. Its warm sector increases in amplitude and at the same time becomes narrower, with the cold front steadily catching up on the warm front until after a while the cold front overtakes the warm front, first near the centre of the low, but later progressively outwards as development continues. Figure 70(*d*) represents a stage about 1 day later than Figure 70(*c*). At the ground the warm air no longer extends as far as the centre of the low; instead, from the tip of the warm sector to the low's centre a new front has formed, known as an *occluded front*, or an *occlusion*. The transformation of a warm sector into an occlusion is known as the *occlusion process*, and a low developing an occlusion is known as an *occluding low*. The point *P* in Figure 70(*d*) where the three fronts meet is known as the *point of occlusion*.

When the occlusion first starts to form, the low is usually deepening and intensifying rapidly so that the pressure gradients become large and it is often at about this time that the strongest winds are produced. The occlusion in Figure 70(d) is a front separating fresh *Pm* air behind it from older *rPm* air ahead of it, and since the *rPm* air is a little warmer than the fresh *Pm* air the occlusion is a cold-type front near the ground with the *Pm* air lying in a wedge beneath the *rPm* air. At the same time, the *Tm* air has been *lifted off the ground*. A vertical cross-section along the line *AB* of Figure 70(d) is given in Figure 72(a) and shows the relative positions of the three air masses. On most occasions the air mass

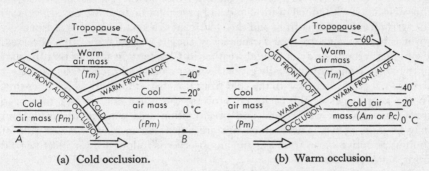

(a) Cold occlusion. (b) Warm occlusion.

FIGURE 72. *Vertical cross-sections through occlusions*

behind the cold front of the warm sector is the coldest of the three, and the occlusion formed is then known as a *cold occlusion*. Sometimes the leading air mass is the coldest, especially in winter when Europe has been flooded with *Am* or *Pc* air, and on such an occasion the surface front is a warm-type front and is known as a *warm occlusion* (Figure 72(b)). Important changes in the movement of the low's centre are found during the stage of rapid deepening just after the occlusion process has started. The centre turns more markedly to the left and at the same time it slows down—see Figure 71.

The final stage in the life-cycle of the low is the rapid occluding of its warm sector, followed by a weakening of the temperature gradients across the occlusion (that is, the front becomes weaker) and a transfer of the low's centre from the tip of the occlusion into the cold air (Figure 70(e)). After this, its movement becomes slow and irregular, pressure rises slowly and it may split into a cluster of several rather indistinct centres near each other.

Whilst the low is occluding, its centre may have moved some 3000 kilometres, say, from the east coast of the United States of America to a position between Iceland and Scotland. Figure 71 represents the positions of our model low at 12-hour intervals shown on a *continuity chart*. The main developments stand out clearly:

(a) increase in amplitude and wavelength of the distortion of the polar front,
(b) occlusion of the warm sector,
(c) curvature of the path of the centre to the left,
(d) variation in speed of the centre.

Most occluding lows reach Britain when the occlusion process is already well advanced; such a low may be represented by Figure 70(d) or (e).

10.1.3 *Weather associated with a warm-sector depression*

We have seen in outline how air masses and fronts are involved during the life-cycle of a warm-sector depression. Now we shall consider the distribution and evolution of the weather during the same period. In Section 10.5.1 we shall find that a developing low is a region of convergence at low levels and divergence aloft, producing widespread upward motion, and cloud formation by adiabatic expansion. The distribution of convergence and divergence is different at each level in the troposphere and this causes a complex pattern of vertical winds. Often the most active regions are near the fronts of the depression which are thus associated with much of the cloud and precipitation in a new low. In general the warm front is an ana front and the cold front is a kata front; this is especially so near the low's centre, but farther away their characters are often reversed. From this and our study of frontal weather in Chapter 9, it follows that the part of the warm front which usually gives the most cloud and precipitation lies near the low's centre, whereas with the cold front it is its trailing part which does this. In the later stages of the occlusion process, the remaining part of the warm front often gives a weak cloud and precipitation area whereas the cold front may be more active. These properties are expressed in a well-known rule of thumb: an active warm front passing an observer is followed by an inactive cold front, and vice versa.

Most of the clouds produced by the widespread, slow upward motion in a new low form within the *warm* air mass, and nearly all of these lie above the frontal zones, that is, from the point of view of an observer on the ground, they lie on the cold sides of the surface positions of the fronts. Near the centre of a low, however, the warm sector air itself is rising slowly, producing an extensive area of cloud which is continuous with the frontal cloud, as shown near point A in Figure 73(d). Also, the cold air near both the low's centre and its fronts may be slowly rising, giving further cloud beneath the frontal cloud. Most other clouds around the low are of the air-mass type. Thus, the Tm air in the warm sector usually has an extensive layer of St or Sc with little cloud above about 1 kilometre. This layer is thickest near the centre and becomes progressively thinner and more broken away from it, that is, towards the adjacent high-pressure area; any drizzle or advection fog is most likely in the northernmost part. The Pm air has its typical Cu and Cb, the latter being found especially to the west of the centre where the north-west winds bring in relatively unmodified Pm air and where low-level convergence accentuates convection. The Pm air ahead of the warm front is usually stable aloft as a result of previous subsidence and often has only $Cu\ hum$ or $Sc\ cugen$.

It remains to describe the weather found with occlusions. When the cold front catches the warm front the two cloud systems combine and they are supplemented by further layers at low levels, especially in the less cold of the two polar air masses. An approaching occlusion is heralded by clouds very similar to those ahead of a warm front. If it is active they would be Ci and Cs, followed by As and Ns with *pannus* below; if less active it might be Ci and Cs, with separate thin layers of As and Ac below. Its difference from a warm front is soon noticed after the front passes, however, because instead of Tm air following it (in a warm sector) there is a direct change to another polar air mass.

The patterns of cloud and precipitation around the low change as it develops; Figure 73(a), (d) and (g) show likely patterns at three stages in the life of our model low. If its centre passes to the north of an observer (a typical situation

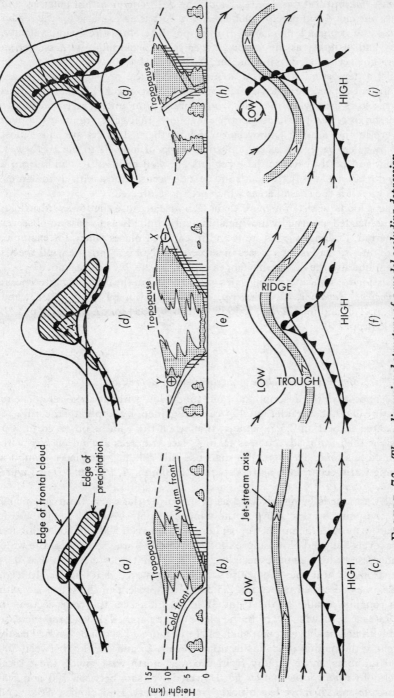

FIGURE 73. *Three-dimensional structure of a model occluding depression*

Top row: horizontal distribution of frontal cloud and precipitation.
Middle row: vertical cross-sections along straight lines shown in top row.
Bottom row: streamlines at 300 millibars.

over Britain) he will experience a sequence such as *Pm—warm front—Tm—cold front—Pm*, but if it passes to the south he will remain within polar air and have only one cloud and precipitation area. It is left as an exercise for the reader to discuss the sequence of changes in temperature, pressure, wind velocity, humidity and visibility as the various stages of an occluding low pass over an observer. The accounts given in Chapter 9 will be useful.

Whilst a depression is in its most intense stage the pressure gradients are often sufficient to produce winds of gale force or even stronger. The most vigorous and intense lows form when the polar front is most pronounced, and since winter is the season of greatest air-mass contrast it follows that gales are most frequent and severe in that season. Our western and northern districts are those most affected by gales, partly because the normal tracks of lows lie to the north-west of Britain, and partly because these regions are well exposed to winds from a westerly direction. Gales in the south and east do occur, almost entirely in winter, but even then it is the coastal areas which are most affected.

A gale seldom persists for more than 24 hours because the lows with which they are associated are usually moving sufficiently quickly as to pass an observer in that period. Prolonged gales are found in exposed places when, for example, a series of intense depressions passes in succession. Not only are the wind speeds large then, but also the direction changes rapidly as the low centres move by.

Discussions of some interesting gales will be found in the following references: *Meteorological Magazine*, **82**, 1953, pp. 71, 97 and 313, **84**, 1955, p. 333, and **85**, 1956, p. 290; and in *Weather*, **7**, 1952, p. 116, **9**, 1954, p. 67, **25**, 1970, p. 129, and **27**, 1972, p. 110.

10.1.4 *Three-dimensional structure of a warm-sector depression*

We cannot leave the description of our model low without some reference to its structure *aloft*. Some idea of this can be gained by applying the rule of thumb stated in Section 2.5.5, namely, that when the wind is observed to *veer* with height then *warm* advection is taking place, whereas a wind *backing* with height indicates *cold* advection. We shall use this rule at four places around a developing warm-sector low, represented by the points *A*, *B*, *C* and *D* in Figure 70(*c*).

At *A* warm air is being advected as is shown by the warm front advancing towards the point, so that above *A* the wind direction *veers* with height and it is veered the most in the upper troposphere, say between 500 and 300 millibars. At these levels the wind blows approximately parallel to the front. Also the wind speed increases with height, especially in the upper troposphere, so that it is usually strongest near the tropopause where it may exceed 100 knots. Indeed, a jet stream as described in Section 2.5.3 is often present in the warm air with its axis roughly parallel to the front. The jet intersects the cross-section in Figure 73(*e*) at *X*, blowing *out* from the plane of the paper. The veer and increase of wind with height above a point such as *A* in Figure 70(*c*) can be verified easily by watching the motion of clouds at upper levels. *Ci* and *Cs*, say, between 500 and 300 millibars, move rapidly from the west-north-west giving the typical streamers of *Ci fib*, *Ci unc* and *Cs fib*. Lower down, say, between 700 and 500 millibars, *Ac* and *As* move less quickly from the west. Low clouds, *Sc str* and *pan*, come from the south-west.

At B cold air is being advected as is shown by the cold front moving away from the point, so that above B the wind *backs* and increases with height to give a jet stream whose axis intersects the cross-section in Figure 73(*e*) at Y, blowing *into* the plane of the paper. We can check these winds too by watching the movement of the clouds. Ci and Cs come rapidly from the south-west, Ac and As less quickly from the west-south-west, and low cloud from the north-west.

At C there is no advection of either colder or warmer air towards the point. The warmest air lies towards the south so the thermal wind (Section 2.5.5) is westerly for all layers in the troposphere—the deeper the layer considered, the stronger is the thermal wind. Now the value of an upper wind can be found by simply adding the lower wind to the thermal wind vectorially. If we take the lower wind here as being the surface wind, which is easterly, and add to it the westerly thermal wind, we get an upper wind which is either easterly or westerly depending upon whether the thermal wind speed is less or greater than the surface wind speed. Since the thermal wind speed increases with the depth of the layer considered, there must be some level at which the direction of the upper wind reverses—below this level it will be easterly, and above it will be westerly.

At D, within the warm air mass, the thermal wind for any layer is small because the air mass is relatively uniform. It is usually westerly since the air mass is likely to be warmest at low latitudes. So, in the warm sector, the wind speed increases slowly with height but changes little in direction from the west.

We have now seen how the wind velocity varies with height at four representative places around a warm-sector depression. These variations, which have been confirmed on innumerable occasions by direct measurement, can now be used to construct the pattern of winds at some fixed upper level over the low, say, at 300 millibars. Figure 73(*f*) shows the chart for 300 millibars with some contours drawn and the positions of the fronts at 1000 millibars indicated. Two features are prominent: a *ridge*, with its axis lying a little to the east of the 1000-millibar position of the warm front, and a *trough*, with its axis behind the cold front. Notice also the absence of a low centre in the contours. At this stage in its development the low has a closed circulation up to only about 700 millibars, but as it gets older so centres develop at progressively higher levels. Thus, 24 hours later the contours in Figure 73(*i*) show a distinct centre. A fully occluded low often has a closed centre at all levels in the troposphere and even above, each centre being roughly in the same place. Notice, too, how the jet stream becomes distorted compared with its shape during the very early stages of the low (as in Figure 73(*c*)).

We have described a model developed by Bjerknes, Bergeron and others, and no individual depression can be expected to conform in all details. Nevertheless, cloud pictures obtained from meteorological satellites (see Appendix B, Section B.2.1) do confirm the broad pattern on many occasions. Figure 74 represents a schematic picture of the cloud systems associated with a family of depressions as seen from a satellite. The arrows represent streamlines of the flow near the ground, and we can see many features in agreement with Figure 73. For example the jet stream is of the same shape and in the same relative position as shown in Figure 73(*i*) and the upper-level trough (BC) agrees well with the trough shown in the 300-millibar streamlines. Open cells indicate active convection with well-developed cumulus and showers in Pm air, while closed cells indicate convection damped down by subsidence in the surface ridge or anticyclone.

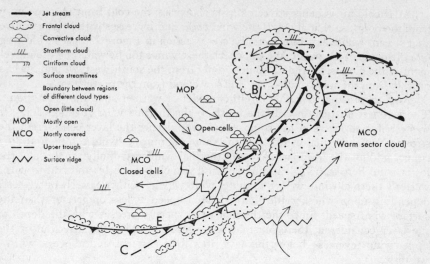

Jet stream
Frontal cloud
Convective cloud
Stratiform cloud
Cirriform cloud
Surface streamlines
Boundary between regions of different cloud types
O Open (little cloud)
MOP Mostly open
MCO Mostly covered
Upper trough
Surface ridge

FIGURE 74. *Schematic representation of cloud patterns associated with a family of depressions*

10.2 OTHER TYPES OF FRONTAL DEPRESSIONS

10.2.1 *Cold-front waves*

Occluding warm-sector lows similar to the model described in Section 10.1 are fairly common features on synoptic charts. However, other types of frontal depressions occur, and a few of them are described here.

The trailing cold front of a partly occluded depression is a favourable region for the growth of new waves, especially if the temperature contrast across the front is large. Waves forming here, known as *cold-front waves*, move quickly along the front in the usual way, often with speeds of 20 to 40 metres per second (40 to 80 knots). Many are stable and dissipate after travelling perhaps 2000 kilometres. Their main effect on the weather is to temporarily slow down, or reverse, the general southward movement of the cold front, and thus also of the encroachment of polar air into lower latitudes. It was often difficult to place cold-front waves accurately on a synoptic chart when they formed over the Atlantic where observations were sparse. However, since satellite cloud pictures became regularly available (see Figure 74, Plates XIX and XX, and Appendix B, Section B.2.1) these waves have presented much less of a problem to the chart analyst in Britain.

Some cold-front waves are unstable and may develop wind circulations comparable in size and intensity with their predecessors. A section of the polar front lying across the North Atlantic, say, may then show two or more lows simultaneously, each evolving to a large extent independently and each at a different stage of development with the easternmost usually being the most advanced. The members of such a *family of depressions* move along similar tracks, but the cold polar air usually pushes farther south behind each low so that the track of any given member is often farther south than that of its forerunner. The series is ended when polar air breaks through to low latitudes and

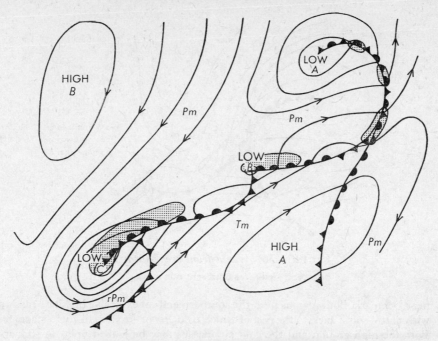

FIGURE 75. *A family of depressions*

Low *A*: old occluded low (Section 10.1.2).
Low *B*: cold-front wave (Section 10.2.1).
Low *C*: slow-moving low at low latitudes (Section 10.2.3).
Also, High *A* is part of the subtropical high-pressure belt and High *B* is a slow-moving
blocking high.
Shaded areas represent rain areas.

the polar front degenerates, only to re-form again at higher latitudes. A family
may consist of two to perhaps five members. Figure 75 shows a situation with a
family of three, each of a different type.

10.2.2 *Break-away depressions*

Waves may also be found on warm fronts but they are more rare than cold-
front waves. They are most likely to develop when the air mass ahead of the
front is particularly cold as, for example, when Europe has been flooded by *Am*
or *Pc* air in winter. A *warm-front wave* travels along the front in the usual way
so that it moves away from the main low-pressure centre; because of this they
are known as *break-away lows*. When a break-away low moves out of the circulation of the original it may draw cold air from the north into its own circulation. This is frequently followed by rapid development and the low may
become an important feature of the chart. Even a relatively small warm-front
break-away low often greatly modifies the weather one would expect to find near
the front. Figure 76 shows a synoptic situation where a warm-front break-away
low is developing.

Break-away lows can also form at the point of occlusion. Two geographical
regions favourable to their formation are the southern tip of Greenland and

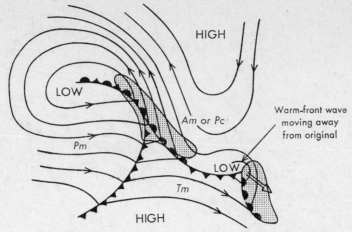

FIGURE 76. *Warm-front break-away low*

Shaded areas represent rain areas.

near Denmark. Both regions lie at the southern ends of extensive highland masses with north–south axes, to the east of which cold *Am* or *Pc* air is likely to stagnate. Both the high ground and the cold air masses may be looked upon as barriers preventing the eastward movement of the occlusion (Figure 77).

10.2.3 *Cut-off depressions*

Examination of upper-air charts over a developing warm-sector depression shows that the approximately west-to-east flow is only temporarily disturbed; this may be seen in Figures 73(*c*), (*f*) and (*i*). The contours usually form a ridge and trough which move quickly in association with the low shown on the surface chart. Sometimes, however, these troughs and ridges are seen to grow in amplitude with their axes lying roughly north to south, and at the same time they become much slower moving. When this occurs

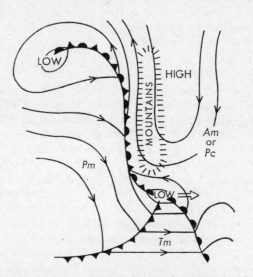

FIGURE 77. *Break-away low at the point of occlusion*

SATELLITE
PLATES

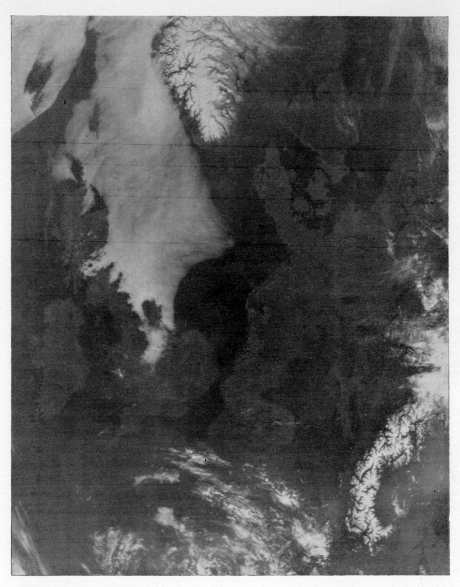

Satellite picture, visible image

Visible image (VIS) from NOAA-3 Very High Resolution Radiometer

PLATE XIX

The IR display is arranged so that the relatively cold clouds, snow and ice are shown in light tones (less radiant energy reaching the radiometer) and warmer surfaces are shown in darker tones (more energy reaching the radiometer). The VIS display is arranged such that highly reflective surfaces such as clouds, snow and ice appear as light areas and low-reflectance, cloud-free areas as darker areas. Note the bright area over most of the North Sea in the VIS; corresponding areas in the IR imagery are relatively dark, indicating that they are rela-

Satellite picture, infrared image

Infrared image (IR) from NOAA-3 Very High Resolution Radiometer

PLATE XX

tively warm surfaces similar in temperature to the clear areas of the southern section of the North Sea between England and the Low Countries. Again, the combined information content of the VIS and IR images suggests that the bright areas in the VIS correspond either to fog or to low stratus. Snow over Scandinavia and the Alps is distinct in both images. *Plates XIX and XX are reproduced by courtesy of the United States Department of Commerce, National Oceanic and Atmospheric Administration.*

a trough is found to be occupied by cold air (and hence shown as a cold trough on a thickness chart) and a ridge by warm air (a warm ridge on the thickness chart).

Because of their size—typical amplitudes being 30° of latitude and wavelengths 90° of longitude—only three to five troughs, with their intervening ridges, can coexist over the whole of the northern hemisphere. Their contour patterns are continuous around the globe in the form of distorted sinusoidal waves. These large upper troughs and ridges are known as *Rossby long-waves*, named after the Swedish meteorologist who first studied them in detail in 1939. They are remarkable for their persistence, perhaps for several days or even weeks, contrasting with the rapid evolution of occluding lows which are on a smaller scale. Because of their large amplitude, cold air is able to flow to quite low latitudes in the troughs, perhaps south of 40°N, and at the same time warm air in the ridges flows well north to 70°N or more. This means that the polar front must lie much more in a north–south direction and at the same time it becomes broken into several portions, each lying on the eastern side of a cold trough.

In its most extreme form a Rossby trough can develop a closed wind circulation at its southern end. This consists of a mass of polar air which has been completely cut off from its source region to the north. On a thickness chart this air is shown as a closed cold centre in the thickness lines, that is, a cold pool. A closed wind circulation is usually present at all levels throughout the troposphere and the mode of origin of this type of low has led to its being called a *cut-off low*. It moves slowly, even westwards on occasions (that is, the reverse of the motion of occluding lows), it is persistent and is centred in relatively low latitudes (usually between 50° and 30°N). Fronts indicating the edges of the cold air are diffuse and difficult to find. Weather typical of a cut-off low is of the polar air-mass type intensified in places by convergence at low levels. Areas of thundery rain or thunderstorms are experienced over southern Britain when such a low is slow moving over western Europe in summer. First signs of this rain are usually shown by *Ac cas* or *Ac flo* as in Plates XIV and XV. Particular geographical areas seem to be favourable to their formation, especially near the south-east coast of the United States of America, between the Azores and Portugal, and over the middle of the Mediterranean Sea.

Not always does a cold pool form at the southern end of a Rossby trough. Occasionally cold air may continue to feed southwards on the western side of a low which has developed at the southern end of that portion of the polar front lying on the eastern side of the upper trough. Such a low is not devoid of fronts; its structure resembles that of an occluding low with a warm occlusion. The *Pm* air on its western side is greatly modified as it rounds the depression in low latitudes, so that when it returns there is little temperature contrast between it and *Tm* air, and the cold front is diffuse. A low of this type is shown as low *C* in Figure 75. The warm front and warm occlusion, however, remain quite definite. These lows are slow moving and associated with widespread and persistent rain. In winter a low of this type which is moving slowly up the English Channel can give heavy snowfall over southern Britain; the air-mass contrast between the polar air on its northern side (usually *Pc* or *Am*) and the *Pm* or *Tm* air to the south is then extremely well marked. A break-away low may form on its warm front.

10.2.4 *Some modifications of simple patterns*

One of the greatest difficulties in analysing synoptic charts for air masses and fronts arises from the fact that many of the surface observations are unrepresentative of the air mass within which they are made. This is through no fault of the observers, but results largely from diurnal variation of the properties of the air mass, or from influences of the topography around the observer. We shall consider a few examples. In winter the continents in middle and high latitudes are often covered by a film of cold air which solar radiation is incapable of removing completely. This film may be only a few hundred metres deep, but it is dense and relatively stagnant, so that fronts penetrating into a continent from the ocean may pass a station which is immersed in the cold layer without an air-mass change occurring at the ground. Aloft, of course, the change is normal. When this occurs the front is said to be *masked*. Conversely, in summer the intense day-time heating is likely to destroy the air-mass contrast near the ground, so that again the effect is to prevent the identification of a clear-cut front.

An interesting feature which may be observed in summer is a temperature *rise* accompanying the passing of a cold front during the day-time. This can occur because the broken cumuliform cloud behind the front allows greater solar radiation to reach the ground than does the complete cover of stratiform cloud ahead of it.

Mountains are notorious for their fickle weather. We have already noted some of their effects on winds (Sections 2.4.2 and 2.6.1), clouds (Chapter 5), precipitation (Chapter 6) and temperature (Section 6.1.3); here we shall consider their influence on fronts. The tendency towards light winds and clear skies in the lee of a mountain barrier allows for considerable modification of any air mass there. For example, in summer this modified air mass tends to be warmer than its surroundings, whilst in winter it tends to be cooler. In this way new and localized air masses are formed. When a front crosses the mountains, the air mass following it will thus meet an air mass in the lee which is different from that in open country and so the front will be modified. To the lee of mountains, warm fronts in summer and cold fronts in winter tend to be weak and diffuse because of lack of air-mass contrast. Conversely, warm fronts in winter and cold fronts in summer tend to be accentuated because of increased air-mass contrast. Similar *diurnal* effects may be seen too, when summer is represented by the day-time and winter by night.

In winter the cold air in the lee of a mountain barrier acts in a similar way to the continental cold air film mentioned above, so that after crossing the mountains a front may not reach the ground until it is well to their lee. Following the movement of a front on a series of synoptic charts may then show an apparent stagnation over the mountains for several hours, followed by a 'jump' far to the leeward. Aloft, the front passes unaltered. Vertical motion of the air masses becomes complex: orographic and frontal ascent are superimposed. On the windward side they supplement each other, but on the leeward side they conflict. Hence frontal precipitation is intensified to windward and decreased to leeward.

An air mass forming a warm sector over Britain is usually *stable*, but occasionally, especially in the summer, it is *unstable*. This is most likely when the air mass is *Tm* which has flowed up from the south and in so doing has been heated by day over France. Alternatively, old modified *Pm* air fills the warm sector. When instability is present it is usually small, but if accompanied by large relative humidities through the whole depth of the troposphere it may give extensive convective clouds at many levels—typically *Cu*, *Cb*, *Ac flo*, *Ac cas* and *Ci spi*—when the air is lifted near the developing low or its fronts. Lines of thunderstorms may then occur and the rainfall is no longer continuous and uniformly spread but becomes sporadic and locally heavy.

10.3 NON-FRONTAL DEPRESSIONS

10.3.1 *Thermal depressions*

Although most lows of middle latitudes develop from distortions of the polar front, others may be found which do not involve any front during their formation. These lows will clearly have weather of an 'air-mass type' only. They are of three types and, although infrequent and usually small, can produce important changes in the weather over Britain.

In Section 2.1.4 we saw that the diurnal variation of atmospheric pressure is a result of the diurnal cycle of solar radiation. Since temperature changes caused by solar radiation are larger over land than over the open sea, it follows that the diurnal pressure changes are also larger over land. Minimum pressure occurs at about 1600 local time so that by this time pressure is lower over land than over the sea. Over Britain the difference is only 1 to 2 millibars and even less in winter. If, during the morning, before heating becomes effective, the isobars over land were widely spaced, the fall in pressure by the afternoon could bring about the formation of closed isobars with a centre of low pressure. Such a low is known as a *thermal low*; it is weak and only a temporary feature because the cooling during the following night causes a pressure rise over land and the low disappears. The best examples are found when solar radiation is strong—during the summer. If solar radiation is weak or the spacing of the undisturbed isobars is closer, a low centre may not form, only a weak *thermal trough*.

Thermal lows and troughs can be found over Europe on most sunny days in summer. The weather associated with them is similar to that of the air mass in which they occur but this, of course, is more unstable as a result of the heating. If the air mass is Tc or Tm the weather may be hot, dry and relatively cloud-free; if it is Pm, showers and thunderstorms are likely.

In tropical latitudes day-time pressure falls in summer often just exceed the following night-time rises, so that pressure inland may fall slowly and progressively from day to day. This leads to a more or less permanent thermal low in summer, the best example of which is the summer low over northern India. The winds blowing around such a low show little daily variation and develop into well-defined streams known as *monsoons*.

10.3.2 *Polar depressions*

The relatively intense heating of the air needed to produce a thermal low is not entirely confined to land areas in summer. When fresh polar air flows away from its source region to cross the open ocean it is rapidly heated from below through contact with the warm water, and pressure falls just as it does over land but with two important differences: firstly, the heating is continuous day and night because the diurnal variation of sea temperature is negligible; secondly, the greatest temperature differences between air and ocean are found in winter. The lows which result from this heating are thus most likely to be found in winter and they are more significant than thermal lows because of their greater intensity. Since they are embedded in a polar air mass they are known as *polar lows*. With weaker heating, only a *polar trough* may form.

A favourable place for the development of polar lows and troughs is in the northerly air stream on the western side of a large occluding low. The cold Pm air streaming southwards is rapidly heated and a series of lows and troughs appear, as illustrated by Figure 78. Added to the normal instability of the Pm air is convergence at low levels, so that the lows and troughs are accompanied by intense convection, giving heavy squally showers, sometimes merging to

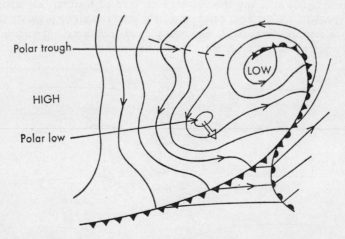

FIGURE 78. *Polar low and trough in the Pm airstream in the rear of an occluding depression*

prolonged outbreaks of rain or snow. Unlike thermal lows, which remain quasi-stationary over land, polar lows drift slowly in the *Pm* airstream, often rather erratically. On moving inland in summer, day-time heating adds further to the instability. This description of polar lows may well be true of many systems but detailed radar studies of a particular low which crossed the United Kingdom showed that it was associated with a low-level (lowest 2 kilometres) thermal contrast in the polar air.

The occurrence of *Cu* and *Cb* clouds, and showers, at times when they would not be expected if they had been the product of diurnal temperature changes, is often the result of low-level convergence in a polar low or trough. Even in winter, or at night, showers may develop inland where otherwise they would be quite unlikely.

In some respects old occluded lows and cut-off lows resemble polar lows since all of them are composed of polar air. The cause of the low pressure in a polar low is the relatively warm sea, so that passage of such a low on to a land surface in winter is usually followed by its filling, but most polar lows show a distinct tendency to remain over the sea. For example, a polar low approaching our south-western districts from the west will tend to move up the English Channel rather than across northern France or southern England. When this occurs continued convection over the sea can give heavy showers near the south coast.

10.3.3 *Orographic depressions*

When a broad uniform airstream flows across a mountain range the straight flow is distorted so that a trough forms just to the leeward (Figure 79); the higher the mountains, the bigger the distortions, so that in the extreme a closed low centre may form, known as an *orographic low*. It is quasi-stationary relative to the mountains and disappears when the airstream ceases to flow. In the lee the downward flow and föhn effect tend to give clear skies so that, in summer, an orographic low is often a region of hot, dry and sunny weather. In winter a layer of cold air may form near the ground allowing widespread fog to develop. Over Europe orographic lows and troughs are well-marked features, especially to the south of the Alps. Our smaller mountains in Britain can give a pressure reduction of up to a few millibars over, for example, eastern Scotland and north-east England when a westerly airstream crosses the country.

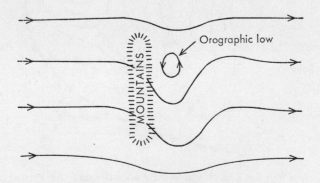

FIGURE 79. *Orographic low in the lee of high ground*

10.4 TROPICAL STORMS

10.4.1 *Introduction*

Although of minor importance as regards British weather, we cannot leave our review of depressions without a brief look at that most violent and destructive of all weather phenomena—the *tropical storm*. It is a feature of low latitudes between the subtropical high-pressure belt and the equator, a region where the trade winds blow throughout the year except where monsoons develop (see Sections 9.2.2 and 10.3.1). The weather of the trade winds is characteristically fine with broken *Cu* whose vertical extent is limited to 2 to 3 kilometres by the 'trade-wind inversion' caused by subsidence in the high-pressure belt. Weak troughs, or *easterly waves*, appear at times and move slowly *westwards*. Convergence in these troughs promotes the growth of large *Cu* or *Cb* and bands of showers develop. On a few occasions these troughs grow rapidly, convection becomes organized into a definite pattern, pressure falls rapidly and a deep and intense tropical storm forms. It is small compared with a depression of temperate latitudes, being only about 300 kilometres in diameter, but its central pressure is as low as 960 to 920 millibars and its strongest winds 50 to 100 knots with gusts much stronger. Central pressures below 900 millibars have been recorded, the extreme being 877 millibars in a Pacific storm on 24 September 1958.

10.4.2 *Structure of a tropical storm*

In recent years much has been learnt about the structure of these storms, Figure 80 showing the main features diagrammatically. The isobars are approximately circular and closely spaced, especially near the centre. The strongest winds are found in a ring between about 15 and 30 kilometres from the centre; they become progressively lighter outwards. At lower levels, below about 3 kilometres, the winds spiral inwards but aloft, especially above 10 kilometres, they spiral outwards.

Clouds are typically convective with belts of *Cu* and *Cb* lying in patterns which clearly demonstrate the inward spiralling winds of low levels. Aloft, the *Cb* tops are carried outwards as *Ns*, *Cs* and *Ci spi cbgen*, the outer fringes being several hundred kilometres from the centre. Heavy, squally showers and thunderstorms occur within a radius of about 150 kilometres; they become increasingly frequent and prolonged as the centre is approached and the heaviest rain is found in a ring roughly just outside the ring of strongest winds.

A characteristic feature of each mature tropical storm is its 'eye', a central region with light winds and broken cloud. It is quite small—about 8–15 kilometres in radius—and is occupied by very warm, subsiding air; its extreme warmth helps to account for the very low surface pressure at the centre.

A tropical storm always develops over the open ocean but only over those parts with a high temperature—usually above 27 °C. However, they have never been observed to form between about 5°N and 5°S, nor near coastlines. Later in its life a tropical storm may drift inland, when rapid decay usually sets in and after 1 or 2 days the most violent weather has gone. Most move westwards at about 10 knots but many later turn north or even north-east. As they move out of tropical latitudes they tend to die out, but sometimes they approach and distort the polar front, eventually becoming transformed into a frontal depression. Their relatively slow speed means that a place lying near the track of a centre will be under its influence for many hours. Rainfall amounts are large, 250 to 500

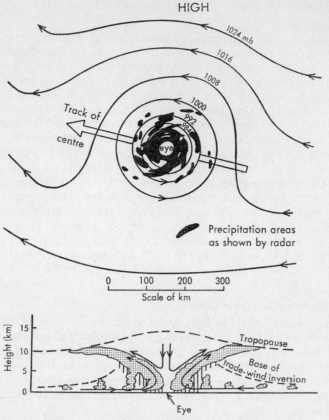

FIGURE 80. *Surface isobars and vertical cross-section illustrating the structure of a tropical storm*

millimetres being not uncommon, so that severe flooding adds to already wide-spread wind damage. Near coasts very high seas are a further danger.

The three best known geographical regions for tropical storms are the southern part of the North Atlantic with the Caribbean Sea, the northern part of the Indian Ocean, and the south-western part of the North Pacific with the China Seas. In these regions they are known as *hurricanes*, *cyclones* and *typhoons* respectively. Late summer is the season of maximum frequency, that is, when sea temperatures in the formation areas are greatest.

On rare occasions, the remains of a hurricane approaches Britain from the south-west. Although it has lost the severe weather of low latitudes it may bring us strong winds and thundery rain.

10.5 MECHANISMS OF DEVELOPMENT OF DEPRESSIONS

10.5.1 *Convergence and divergence*

So far in this chapter our survey of depressions has been almost entirely descriptive. In this last section we shall examine some of the mechanisms which help to explain the development of a depression. The reasons for its initial formation are still obscure so that the problem of why a depression should appear when and where it does is still largely unanswered. Apart from this, recent advances in the study of the motion of the atmosphere within pressure systems has enabled us to see something of the processes causing a continuation of development once it has started.

A depression forms when atmospheric pressure falls over a limited area of the earth's surface, and such a fall is brought about by the removal of some of the atmosphere overlying that area. Similarly, a pressure rise results from the accumulation of air above the area. Enormous masses of air are involved in these changes. Suppose the pressure over a circular area of radius 1000 kilometres (equal to that of a large depression) falls by 1 millibar. We shall calculate how much air must be removed to do this. Now 1 millibar is equal to a force of 100 newtons (that is 100 kg m s^{-2}) pressing on an area of 1 square metre, so by taking $g = 9 \cdot 81$ m s^{-2}, we see that 1 millibar would be produced by a mass of approximately 10 kilograms (strictly 100/9·81 kg) pressing on an area of 1 square metre. So a change of 1 millibar would be caused by a decrease in the mass of the atmosphere of about 10 kilograms over each square metre or $10 \times \pi \times 10^{12} = 10^{13} \pi$ kg $\approx 3 \times 10^{10}$ tonnes over a circular area of radius 1000 kilometres. A change of only 1 millibar over such a large area is quite small; changes of 10 millibars or more sometimes occur.

Consider the horizontal wind flow at some level above the area of development. If *divergence* (see Section 2.6.3) is occurring in this wind flow, air will tend to be removed from that level so that pressure will tend to fall at all levels in the atmosphere below the one under consideration, even down to the ground. Hence, divergence aloft can (but not always will) produce falling pressure at the ground; conversely, convergence aloft usually produces rising pressure at the ground.

Wind flow which is either convergent or divergent is never balanced—individual air parcels are accelerating—so that the wind cannot be geostrophic. In Section 2.4.1 we saw that the observed wind, away from the effects of friction near the ground, is always nearly geostrophic. The difference between the observed and geostrophic wind is small but significant, and occasionally very significant, so now we must look for some causes of acceleration in wind flow. Without these accelerations, when all winds would be exactly geostrophic, there would be no convergence and divergence and therefore little vertical motion and weather.

10.5.2 *Origin of convergence and divergence*

Accelerations in wind flow may be brought about in many ways, but we shall consider only two special situations, in each of which the acceleration of an air parcel is in the same direction as its motion.

(a) Air flowing along a path with decreasing cyclonic curvature downwind (or increasing anticyclonic curvature) is divergent; air flowing along a path with increasing cyclonic curvature downwind (or decreasing anticyclonic curvature) is convergent.

Consider the parcel in Figure 81 moving from *A* to *C* parallel to the curved isobars which are parallel themselves and which have a curvature decreasing from *A* to *C*,

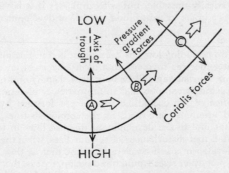

FIGURE 81. *Flow between parallel isobars whose cyclonic curvature decreases downwind*

where it is zero. In Section 2.4.2 we saw that the gradient wind must have an acceleration towards the centre of curvature of its path, so the pressure gradient force and the Coriolis force cannot balance. For a cyclonically curved path the former force is the greater and the difference between them increases with the curvature of the isobars. At point A the curvature is greater than at B; hence the acceleration towards the instantaneous centre of curvature is greater at A than at B so that, although the pressure gradient force exceeds the Coriolis force at both places, the difference is greater at A than at B. At point C, where the isobars are straight, the two forces are equal. Now the pressure gradient force is the same at all three places (the isobars are parallel) so it follows that the Coriolis force must increase from A to C, which means that the parcel's speed must also increase from A to C. Parcels thus accelerate when moving from A to C, and are divergent. Hence, air flowing along a path with decreasing cyclonic curvature downwind is divergent. The other possibilities can be proved similarly. From Figure 81 it can be seen that a region of divergence exists ahead of the axis of a trough; if the trough moves so does this region of divergence. A slowly advancing trough in the upper troposphere has a region of divergence ahead of it and consequently falling pressure at the ground. Here we have assumed that the speed of the parcel of air is much greater than that of the trough and so the air is flowing through the system.

(b) Air in an isallobaric high is divergent; in an isallobaric low it is convergent. Consider an area with uniform surface pressure. If pressure is rising everywhere, but most rapidly at the centre, then an isallobaric high (see Section 2.1.4) covers the area. Initially there is no wind because there is no gradient of pressure. After a while a set of closed isobars forms, with highest pressure in the middle. The appearance of these pressure gradients causes the air to move, accelerating outwards across the isobars away from the newly formed high-pressure centre—the air is diverging from the isallobaric high. It will continue to do this as long as the pressure continues to change in the same way. Conversely, an isallobaric low is a region of convergence. The same effects would be produced even if pressure gradients were present initially.

Most surface lows seem to form as a result of divergence aloft, the first effect being the generation of an isallobaric low at the ground. This in turn produces convergence near the ground which tends to accumulate air in the development area, so raising the surface pressure. We are now faced with a dilemma: the divergence aloft promotes the formation of a surface low whereas convergence below, itself a direct consequence of the low's formation, tends to destroy it. Clearly, the low will continue to develop only as long as the upper divergence is more effective in removing air than the lower convergence is in replacing it. If the divergence were to continue unchecked then absurdly low pressures would result. The convergence below largely compensates for the divergence aloft (the so-called 'Dines' compensation') and ensures that the net rate of fall of pressure at the ground is relatively small.

A very important consequence of this distribution of convergence and divergence in a developing depression is the upward motion of the troposphere, slowly transferring air near the ground to higher levels in the troposphere. This vertical flow will occur most easily if the air in statically unstable, but with difficulty if it is very stable. Stability of the atmosphere acts as a brake by controlling the rate of development of the low.

10.5.3 *Baroclinic zones*

The life-cycle of an occluding low has been described in Section 10.1.2; we shall now consider some of the processes involved in this life-cycle in an attempt to explain its observed development. Our survey will be only very sketchy and incomplete because the whole problem is exceedingly complex and still imperfectly understood.

After the *description* of the occluding process put forward by the Norwegians had been widely accepted, attempts were made by meteorologists to *explain* theoretically why fronts were so fundamental to the development of most lows in temperate latitudes. In 1930 Solberg calculated that disturbances of certain sizes, when they appeared on a front, would continue to develop in a self-accelerating way, that is, they would be *dynamically unstable*. Disturbances of other sizes would not develop, that is, they would be *dynamically stable*. The calculated sizes compared very reasonably with those observed.

Solberg based his ideas on theoretical fronts which were taken as simple *discontinuities* between neighbouring air masses. Real fronts, as we have seen, are not discontinuities—

they always have *zones of mixing* perhaps 100 to 200 kilometres wide (see Section 9.6.1). This unreality in the theory was removed later by other investigators when they found that the development of existing pressure patterns occurred only in certain areas with a characteristic property. In such areas it was found that the surfaces of constant pressure in the atmosphere (isobaric surfaces) intersect the surfaces of constant density—and hence also those of constant temperature (isothermal surfaces). A part of the atmosphere where these surfaces cut each other is known as a *baroclinic zone*.

Now a front lies in a well-marked baroclinic zone because near it the two sets of surfaces intersect at a significantly large angle (although only about $\frac{1}{2}°$). This angle is large because in a frontal zone the horizontal gradients of temperature are large and so the *isothermal* surfaces slope at about $\frac{1}{2}°$ to the horizontal. On the other hand, *isobaric* surfaces are almost parallel to the ground. Away from frontal zones, that is, within the more nearly uniform air masses, isothermal surfaces are also nearly horizontal and therefore hardly intersect the isobaric surfaces.

The appearance of a wave-shaped disturbance in the baroclinic zone creates a wind system which tends to increase the disturbance. This leads to a *self-accelerating* process —once the distortion appears it tends to increase progressively.

In 1955, Petterssen suggested that the original disturbance is often produced by an isallobaric low at the ground being superimposed on a front, the isallobaric low itself being caused by a slowly advancing trough aloft.

10.5.4 *Origins of deepening and filling*

Figure 82 shows diagrammatically how the central pressure of an occluding depression might vary with time. We may distinguish three stages in its development:

(a) *growth*, when the rate of fall of central pressure increases with time,

(b) *maturity*, when the rate of fall of central pressure decreases with time,

(c) *decay*, when the central pressure rises.

These changes are produced by variations in space and time of the extent and intensity of divergence and convergence within the depression, and they are still imperfectly understood. During the *growth stage* the divergence aloft (causing removal of air) exceeds the convergence below (causing addition of air) with the result that pressure falls. This fall produces a trough in the surface isobars, and eventually a centre appears with closed isobars. As the surface isobars become distorted so do the surface winds, and they in turn distort the front—that part ahead of the low moves northwards in the southerly winds developing there whilst the part behind the low moves southwards in the northerly winds. In this way a wave-shaped distortion appears which runs along the front with a speed approximately equal to that of the warm air.

If the wave is dynamically unstable it develops with progressively increasing distortion of the front so that eventually the occlusion process starts. Near the centre of the low the wind circulation so greatly distorts the occluded front that it becomes impossible to find, and the centre migrates from the tip of the occlusion into the cold air. There may be a

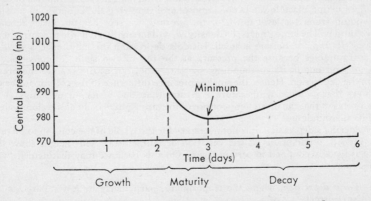

FIGURE 82. *Typical variation with time of the pressure at the centre of an occluding depression*

temporary stage when two centres are present—one on the occlusion and one in the cold air; alternatively, there may be a strong trough lying along the occlusion outwards from the low in the cold air.

Now we have seen that in a developing low there is a slow but widespread ascent of the atmosphere, so that the cyclonic wind circulation which is developing in the lowest part of the troposphere will be slowly carried upwards with the result that a cyclonic circulation appears at progressively higher levels. This upward growth of the circulation introduces a complication: it distorts the wind flow aloft and hence alters the extent and distribution of divergence there. At first it seems to enhance the divergence, with the result that surface pressure falls even faster, but later the changes in the wind flow aloft are such that the divergence decreases rapidly and surface pressure falls much more slowly until finally there is negligible divergence at all levels and surface pressure then stops falling. By this time we reach the *mature stage* when the circulation at all levels has been greatly modified, and we often find a low centre present throughout the troposphere and even above. If the changes do not proceed so far as to give centres aloft, then at least troughs appear.

After the occlusion process is well advanced the central pressure starts to rise, that is, the low fills. There are three important ways by which this can occur: firstly, by the development of convergence aloft above the surface low; secondly, by direct radiational cooling of the atmosphere with its accompanying contraction and slow convergence; thirdly, by friction near the ground causing a slow inflow of air and its accumulation near the centre. The last two are slow processes, but the first can be rapid when the eastern side of a developing ridge aloft advances over the surface low. This occurs particularly when another low is developing just upstream from the first one.

We can see from these ideas something of the causes of the three stages in the life of an occluding low:

(a) *Growth.* Development is self-accelerating and the circulation is largely confined to the lower troposphere. Fronts are important features.

(b) *Maturity.* Development decreases rapidly as the circulation builds up to higher levels. Fronts are of lesser importance.

(c) *Decay.* Pressure rises as the result of convergence. The first two stages whilst the low is deepening are rapid, taking only 2 or 3 days to complete; the last stage is usually slower, taking 3 to 6 days.

An interesting accompaniment to the formation of a depression is an increase in the temperature of the atmosphere (meaned over a great depth) within the area of formation. This can be deduced by using equation (1) of Section 2.1.2:

$$p = \rho g h.$$

Now it is observed that even the most vigorous systems do not extend above a level near the middle stratosphere; higher still, pressure and wind patterns remain almost undisturbed by changes taking place lower in the atmosphere. Applying the equation to a column of air extending from this level down to the ground, we see that the mean temperature of the column will be large (that is, its density, ρ, will be small) when the pressure difference, p, between its top and bottom is small. Inside a depression this difference is smaller than in its surroundings because the pressure at the top of two such columns—one in the depression and one in its surroundings—is the same, but at the ground the pressure is lower in the depression. Hence, the smaller pressure difference of the column in the depression shows that its mean temperature is greater than that of its surroundings. A similar argument shows that the mean temperature of the atmosphere in an anticyclone is less than that of its surroundings.

Considering a depression, it is not necessary for the whole of the column to be relatively warm. Indeed, observations of the variation of temperature with height show that the warmth is usually confined to certain layers. On this basis we may distinguish two types of depression:

(a) *Warm depression*, where the troposphere, particularly its lower parts, is warmer than normal.

(b) *Cold depression*, where the stratosphere is warm whereas the troposphere is colder than normal.

The names are based upon the temperatures of the troposphere. Typical warm depressions are thermal lows, tropical depressions and warm-sector depressions; their cyclonic circulations are largely confined to the lowest troposphere. The fully occluded low is an example of a cold depression; its circulation can sometimes extend from the ground upwards into the stratosphere. In Section 1.3.2 we saw that a warm stratosphere is accompanied by a low tropopause, so a cold depression typically has a low tropopause. The lowest heights of the tropopause found over Britain (below 8 kilometres) are usually associated with centres of large and well occluded depressions.

10.5.5 *Movement of an occluding low*

Both observation and theory show that a disturbance which develops in a baroclinic zone moves approximately with the velocity of the thermal wind between 1000 and 500 millibars through its centre. This is the so-called 'principle of thermal steering'. In the first stage of a low, when the distortion of the polar front is small, the direction of the thermal wind is almost the same as the direction of the warm-sector isobars. The subsequent rapid development causes rapid distortion of the front so that warm air tends to flow northwards ahead of the low and cold air southwards behind it. A consequence of this distortion is that the thermal wind over the centre backs, and so the low turns off to the left of its path. The simultaneous occurrence of rapid deepening with a turning to the left is frequently observed with the mature stage of an occluding low.

When the distortion of the thickness lines reaches an advanced stage, that is, when occlusion has been taking place, little air-mass contrast remains at the centre, so the thermal wind there decreases and the low becomes slow moving. The low-pressure centre is now wholly within cold air and since this is usually coldest to the west, that is, where it is freshest, the light thermal wind present over the centre is from the south; hence the low drifts slowly northwards.

Note that any low which has a well-marked front passing through its centre must move quickly whereas a cold low, with few or no thickness lines through its centre, is slow-moving or quasi-stationary.

10.5.6 *Outline of the development of an occluding depression*

The processes discussed in this section will now be summarized into a tentative outline of the sequence of events during the life-cycle of an occluding low.

(a) An upper trough advances eastwards with a region of falling surface pressure ahead of it.

(b) This region becomes superimposed on a baroclinic zone, especially a front, causing a local distortion.

(c) The distortion develops acceleratively with rapid deepening, a progressive upward extension of the circulation, and a turning to the left of its track.

(d) After 2 or 3 days, pressure ceases to fall and the circulation at all levels reaches its maximum development.

(e) A final slow filling ensues over 3 to 6 days caused by slow convergence.

The fundamental importance of baroclinicity to the occurrence of development is emphasized by the seasonal variation of intensity of frontal lows. They are best developed in winter when the air-mass contrast is greatest.

BIBLIOGRAPHY

BIELINSKI, L.S. and MOONEY, T.W.; 1964. The eye of hurricane Arlene as viewed by radar, 9 August 1963. *Weather*, **19**, pp. 247–248.

BOYDEN, C.J.; 1963. Jet streams in relation to fronts and the flow at low levels. *Met Mag*, **92**, pp. 319–328.

BROWNING, K.A. and HARROLD, T.W.; 1969. Air motion and precipitation growth in a wave depression. *Q J R Met Soc*, **95**, pp. 288–309.

BROWNING, K.A., HARDMAN, M.E., HARROLD, T.W. and PARDOE, C.W.; 1973. The structure of rainbands within a mid-latitude depression. *Q J R Met Soc*, **99**, pp. 215–231.

HARROLD, T.W. and BROWNING, K.A.; 1969. The polar low as a baroclinic disturbance. *Q J R Met Soc*, **95**, pp. 710–723.

JONES, D.C.E.; 1962. Formation of waves on warm fronts in the vicinity of the British Isles. *Met Mag*, **91**, pp. 297–304.

LYALL, I.T.; 1972. The polar low over Britain. *Weather*, **27**, pp. 378–390.

NANCOO, M.E.; 1962. Hurricane 'Hattie'. *Weather*, **17**, pp. 295–304.

PETTERSSEN, S. and SMEBYE, S.J.; 1971. On the development of extratropical cyclones. *Q J R Met Soc*, **97**, pp. 457–482.

WALLINGTON, C.E.; 1969. Depressions as moving vortices. *Weather*, **24**, pp. 42–52.

CHAPTER 11

ANTICYCLONES

11.1 INTRODUCTION

11.1.1 *Some general properties of anticyclones*

An *anticyclone*, or *high*, is characterized by its centre of high pressure around which are roughly concentric isobars and by its clockwise circulation of winds (in the northern hemisphere). Just as lows are three-dimensional features of the atmosphere, so are highs, and their structures vary with height.

Apart from their pressure and wind patterns, highs and lows show other important differences. In general, highs are larger, slower moving and more persistent than lows; their pressure gradients are weaker too, especially near their centres, sometimes so much weaker that a true pressure centre is difficult to find; even several small centres may coexist. These weak pressure gradients make the winds characteristically light and variable near the centre. Whereas a low is usually an independent feature throughout its life of several days, a new high almost always forms either as an extension of an existing high or as a centre near the original, eventually replacing it.

Probably the most important characteristic of a developing anticyclone, and one which largely determines the weather found in it, is the widespread slow descent of air, especially in the middle troposphere, known as *subsidence*. This subsidence has two effects:

(a) It warms the air since descent causes adiabatic compression, and a descending parcel warms at the dry adiabatic lapse rate of 10 °C per kilometre as long as no clouds are present. Since the original environment lapse rate is almost always less than the dry adiabatic lapse rate, it follows that a parcel at A in Figure 83 whose temperature equals that of its surroundings, on descending to a level where its temperature is represented by B, will become warmer than the air originally at that level, represented by C.

(b) It reduces the relative humidity of the air, since the dew-point of descending unsaturated air increases by only about 1.7 °C for each kilometre of descent (see Section 3.2.2) whereas the air temperature increases by

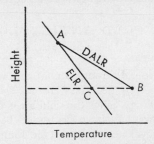

FIGURE 83. *Warming as a result of subsidence*

10 °C per kilometre. Air which has subsided several kilometres has a very low relative humidity.

11.1.2 *Origin of the high pressure in an anticyclone*

In Section 10.5.4 we saw that a depression forms when divergence aloft removes air faster than it can be replaced by convergence at low levels. An anticyclone forms when convergence aloft adds air faster than it can be removed by divergence below. Just as the formation of a depression is accompanied by a rise in the mean temperature of the atmosphere up to a level in the middle stratosphere, so the formation of an anticyclone is accompanied by cooling.

When the distribution of temperature with height in an anticyclone is measured, for example, by a radiosonde, it is found that, in general, it is not the whole of the atmosphere which is colder than normal. The cold air tends to occur in either of two definite layers, which enables us to distinguish two types of anticyclones:

(a) *Cold anticyclone*, where the cold dense air is largely confined to the lower troposphere up to, say, three kilometres. Above this level, temperatures are near normal.

(b) *Warm anticyclone*, where the cold dense air is largely confined to the upper troposphere and the lower stratosphere. The middle and lower troposphere are often warmer than normal.

These differences are shown in Figure 84. The terms cold and warm high refer, of course, to the lower troposphere and are survivals from earlier in this century when temperature observations were confined to that layer.

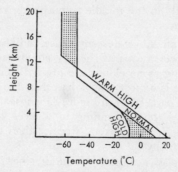

FIGURE 84. *Distribution of temperature with height in anticyclones*
Shaded areas represent layers of cold air.

11.2 COLD ANTICYCLONES

11.2.1 *Structure and properties*

Ascending from the ground through a cold anticyclone we find the depth of cold air overhead progressively decreases and so, therefore, does the excess of pressure over its surroundings. Since the cold air is confined to the lowest part of the atmosphere it follows that above the top of the cold air there is no excess pressure, that is, an anticyclone would not be found on upper-air charts for levels above the top of the cold air. It follows, too, that the intensity of a cold

high is greatest at low levels; hence the clockwise wind circulation is most pronounced there too. A cold high is a shallow feature of the atmosphere.

In our brief discussion of the general circulation in Section 9.2 we saw that
the polar regions above about 70°N have relatively high atmospheric pressure.
This is shown on charts of annual mean pressure as the *polar anticyclone*. In
winter this anticyclone extends well south into the interiors of the northern
continents, and separate centres may be present over North America, Greenland
or Siberia. In summer, only the Arctic Ocean is cold enough to allow a high to
form; even then it is feeble and at times may be absent. Within these highs the
winter sky is often cloudless and temperatures are very low, −20 to −50 °C;
the summer is warmer with temperatures above 0 °C, but advection fog or low
St cloud is common over the sea and coasts.

Over Britain, high pressure commonly accompanies the polar air which
sweeps across the country behind a southward-moving cold front. A *centre* of
high pressure is not common, however, and usually only a *ridge* appears, illustrated
by Figure 86(*a*). The exact shape and movement of such a ridge depends largely
upon the shapes and movements of its neighbouring lows. Its weather is typical
of polar air but is often modified by an inversion which develops at the base of the
subsiding air, separating it from the unsubsided air beneath. This inversion
restricts the vertical extent of convection to, say, two kilometres so that clouds
are of the *Cu* and *Sc cugen* types with normal diurnal changes. Plate IX shows some
of these clouds. Such a ridge moves quickly across Britain, giving a period of
clearer weather for a day to two between neighbouring lows. At night, light winds
and clear skies lead to low temperatures, especially when the ground is snow-
covered in winter. Local radiation fog patches occur but are not usually extensive.
As a ridge moves away it is followed by increasing high and medium cloud ahead
of the next depression.

11.3 WARM ANTICYCLONES

11.3.1 *Structure and properties*

A warm high shown on a synoptic chart is accompanied by correspondingly
high pressure on all upper-level charts up to well into the stratosphere. This
high pressure may be represented on a contour chart as a high *centre* in the
contour lines, but more commonly by a large *ridge*. The anticyclones forming
the chain of centres in the subtropical high-pressure belt are typical warm highs.
Over the North Atlantic the main centre is often found near the Azores. The
Azores high is slow moving, sometimes remaining in the same area for several
days or even weeks, accompanied by a vast area of quiet weather with broken
Cu and *Sc* clouds. On its northern side the westerly winds cross Britain, with
waves and depressions on the polar front and ridges between them, so giving us
typically changeable weather.

Occasionally the Azores high is displaced over western or central Europe,
often as an accompaniment to the development of a deep depression in the
northern part of the North Atlantic. Most of Britain is then covered by *Tm* air,
and the polar front with its disturbances is displaced farther north than normal.
Weather over Britain is of the type described in Sections 9.4.3 and 9.5.3. If
subsidence within the high extends to very near the ground, giving widespread

clear skies, it may, together with the normally light winds and high surface dew-points in the Tm air, lead to the formation of extensive radiation fog, trapped in a film of cold air near the ground, often only a few hundred feet deep. In winter this fog can be very persistent because a warm high, once it is established, tends to persist with little movement and its light winds are not favourable to the dispersal of fog. In summer, day-time heating is usually sufficient to remove the cold film with its fog to give clear skies and high temperatures. Most of our spells of warm, dry weather in summer are found to accompany slow-moving warm highs. In the subsided air aloft relative humidities may be very small, and observers in mountainous regions who happen to be immersed in this air then report relative humidities of 10 to 20 per cent or even less.

Figure 84 shows that the tropopause in a warm high is found at higher levels than normal. The highest tropopauses over Britain (above 12 kilometres) are always associated with well-developed warm highs, and so are the lowest stratospheric temperatures (-60 to -80 °C).

11.3.2 Blocking anticyclones

The classical picture of the general circulation is the one described in Section 9.2, where the belts of high and low pressure lie approximately east to west and where consequently most winds have their greater components parallel to the equator. Such a wind flow is said to be zonal, and in it the wind components perpendicular to the equator are localized or temporary as may be found, for example, ahead of and to the rear of depressions.

At irregular intervals the zonal flow tends to break down with the result that wind components perpendicular to the equator become more important; when this occurs the flow is said to be meridional. A feature of meridional flow is the appearance of one or more large warm anticyclones in the midst of the temperate low-pressure belt, with their centres between about 50° and 60°N. Such a high prevents the polar front from lying in its normal position so that frontal depressions must either move into very high latitudes (70°N or more) or into low latitudes (30° to 40°N). The normal tracks of these depressions are blocked and this has led to the use of the name blocking high to describe such an anticyclone. When a blocking high is present over Europe the Azores high is usually absent.

There is a decided tendency for blocking highs to persist in preferred geographical regions, especially between 10° and 20°W over the North Atlantic, and between 10° and 20°E over Scandinavia. Weather over Britain is greatly influenced by the position of the high. We shall consider some of the associated synoptic situations.

(a) Blocking high over the North Atlantic. Britain is covered by an air mass which has crossed the ocean at high latitudes and approaches us from the north-west (Figure 85(a)). This air mass has properties intermediate between those of Pm and Tm air. It may be old polar air which has circulated around the high before returning to us, or it may be old tropical air returning southwards again. Above about two kilometres it is well subsided so that most cloud is confined to the lowest layer. Inland the cloud normally takes the form of an extensive sheet of $Sc\ str$, often thick and giving a persistent gloom, but over the sea and exposed coasts, and also inland in summer, shallow convection may occur giving slight showers from $Cu\ med$

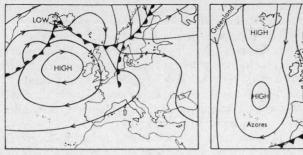

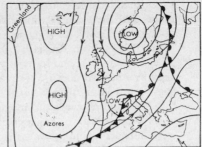

(a) High over the North Atlantic.　　(b) High from Azores to Greenland.

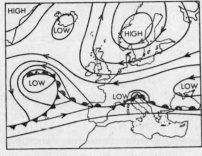

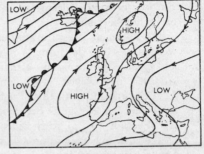

(c) High over Scandinavia.　　(d) High from Azores to Scandinavia.

FIGURE 85. *Typical synoptic situations when weather over Britain is dominated by a blocking anticyclone*

and *Sc cugen*. Much of the *Sc* inland may have its origin in the spreading of these *Cu* tops; in winter the accumulated *Sc cugen* is very persistent and can spread from over the sea to places well inland. The polar front is weak and lies roughly to the north of Britain; if it moves south into Britain the colder air to its north brings clearer skies.

(b) Long blocking high with a north–south axis from the Azores to Greenland. This high usually appears in one of two ways: either as a link-up between a northward extension of the Azores high and a centre of the polar high over Greenland, or as an intensification and southward extension of the Greenland high itself. In both situations a deep low is usually present over Scandinavia so that a stream of *Am* air flows across Britain from the north (Figure 85(*b*)). The weather is typical of *Am* air as discussed in Sections 9.4.1 and 9.5.1. Night minimum temperatures can be very low in winter. The polar front is well marked but displaced to low latitudes, often over the Mediterranean Sea which then experiences particularly stormy weather.

(c) Blocking high over Scandinavia. This is a common synoptic situation, the high often becoming large and intense, especially in winter when on occasions it may extend from Greenland to Siberia. Its development at relatively high latitudes allows the formation in winter of a very cold layer of air near the ground so that, in effect, the high has properties common to both warm and cold highs. Because of this it may develop exceptionally large central pressures, say, 1050 to 1060 millibars or even more, and the easterly winds which blow on its southern side may be strong as a result of the large pressure gradients. These east winds reach Britain from Russia by crossing northern Europe (Figure 85(c)). The air mass is Pc and feels very cold, not only because its temperatures are low but also because they are accompanied by strong winds. The day-time maximum may not exceed 0 °C, an event largely confined to this situation (however, a maximum not exceeding 0 °C may also be observed in Am air or in persistent radiation fog).

The polar front is well marked but at low latitudes. Depressions approaching Britain from the south-west often stagnate and only affect our extreme south-western districts. Sometimes, however, a low moves eastwards up the English Channel or across northern France into central Europe, causing widespread snow, with drifting in the easterly winds on its northern side.

In summer, pressure gradients around the Scandinavian high are weaker, and also the air mass over Britain is Tc so that high temperatures and clear skies are usual—a marked contrast to the winter. In fact our summer 'heat-waves', with temperatures perhaps in the range 27 to 32 °C, are found with this situation. At the same time much of central and southern Europe is covered by a cut-off cold pool of old polar air (see Section 10.2.3) with widespread thundery weather. As the high weakens, a low may approach western Europe from the Atlantic and between them the southerly winds sometimes spread this old polar air northwards across Britain accompanied by thundery outbreaks. It is well known that our hot spells in summer are often ended by thunder.

(d) Long blocking high with a north-east–south-west axis from the Azores to Scandinavia. The axis crosses northern Britain so that in the south there are north-east winds, usually of Pm air which has passed around the north-eastern end of the high over Scandinavia (Figure 85(d)). Weak troughs moving south-westwards in this stream can give thunder in summer, and showers of rain, snow or hail in winter. The polar front lies well out in the North Atlantic, with frontal lows moving north-eastwards across Iceland to northern Scandinavia.

11.4 DEVELOPMENT OF ANTICYCLONES

11.4.1 *Some general ideas*

The pressure rise accompanying the formation of a new anticyclone is the result of an accumulation of air brought about by convergence. Just as a new low seems almost always to form as a result of divergence aloft (with convergence below which is not quite capable of balancing it), so a new high seems to form as a result of convergence aloft. Divergence below will also occur but it too will not be quite capable of balancing the convergence, that is, again Dines' compensation is not quite complete (see also Section 10.5.2). A natural consequence of this convergence aloft and divergence below is the widespread downward motion in the troposphere, or subsidence.

When a cold high is forming, the origin of the high-level convergence is easy to find —it is a result of the contraction of the lower troposphere whilst it is being cooled by contact with a cold ground. We may contrast this with the formation of thermal lows by divergence aloft as a result of the expansion of the lower troposphere whilst it is being warmed by contact with a warm ground. However, the source of the high-level convergence when a warm high is forming is doubtful; originally it may occur just ahead of a slow-moving upper ridge—contrast this with the divergence ahead of an upper trough discussed in Section 10.5.2.

11.4.2 *Development of a warm high*

The development of a typical warm high on the surface chart is accompanied by changes higher in the atmosphere. When a zonal circulation breaks down with the formation of large north–south distortions in the contours, that is, when Rossby long-waves form (see Section 10.2.3), then surface pressure is found to be high beneath the upper ridge, but it is low beneath the upper trough. As we saw in Section 10.2.3, a trough is filled with cold air which at its southern end may become detached as a cold pool and shown on the surface chart as a cut-off depression. Similarly, a ridge is filled with warm air which at its northern end may also become detached as a warm pool and shown at the surface as a warm high. When a trough-and-ridge pair change in this way simultaneously the final state is a cold low embedded in the subtropical high-pressure belt and an accompanying warm high embedded in the temperate low-pressure belt.

This is a reversal of the normal situation; it may be accompanied by a reversal of the normal direction of motion of the centres since each may sometimes be seen to drift slowly *westwards*. Even if their motion were normal it would still be slow and irregular; this is to be expected from the principle of thermal steering mentioned in Section 10.5.5 because each centre is filled with air having an approximately uniform temperature and so each therefore has only a very weak thermal wind crossing it. Once formed, the whole system seems very stable—dynamically stable—and persistent so that the overall effect is to bring a lengthy period of quiet weather to the normally disturbed temperate latitudes whilst changeable conditions spread unusually far south into subtropical latitudes.

A new warm high may be seen quite often to develop from a comparatively minor ridge in the polar air between two frontal depressions. This is because the ridge in the surface isobars underlies the region just ahead of the upper ridge in the contours which develops ahead of the warm front of the westernmost of the two lows. This upper ridge, together with its growing surface ridge, is found to develop most when the western low develops rapidly but without much movement. The surface ridge grows, intensifies and develops a closed centre in its isobars, all as an accompaniment to the growth of the low upstream. Two stages in its life are shown in Figure 86.

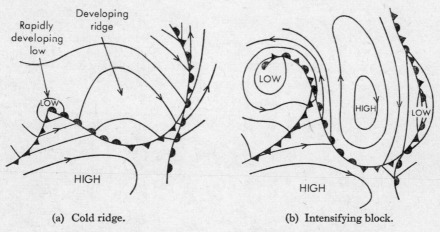

(a) Cold ridge.   (b) Intensifying block.

FIGURE 86. *Growth of a warm high from a cold ridge*

Initially the ridge is a shallow feature because it is produced by the cold air and this lies only beneath the frontal zone. Later, as the ridge aloft develops, it changes into a much deeper feature extending from near the ground up into the stratosphere. Whilst still a ridge it moves steadily eastwards, steered by the thermal wind which itself is the result of the contrast between cold air to the north and warm air to the south. By the time the high is fully developed the contrast between the polar and tropical air masses has become very indistinct because both have been warmed adiabatically as subsidence continues. This decrease in contrast decreases the thermal wind over the high's centre to near zero so that it becomes very slow moving.

BIBLIOGRAPHY

FINDLATER, J.; 1961. Thermal structure of the lower layers of anticyclones. *Q J R Met Soc*, **87**, pp. 513–522.

FINDLATER, J.; 1967. A note on analysis of anticyclones. *Met Mag*, **96**, pp. 69–73.

SUMNER, E.J.; 1954. A study of blocking in the Atlantic-European sector of the northern hemisphere. *Q J R Met Soc*, **80**, pp. 402–416.

CHAPTER 12

FORECASTING

12.1 INTRODUCTION

12.1.1 *Nature of the problem*

The greatest stimulus to the study of meteorology has undoubtedly been the desire to forecast changes of the weather. In principle the problem is very simple, for all we need to know are, firstly, the state of the atmosphere at any given moment and, secondly, the physical laws which govern the changes of that state. When an answer is sought in practice, however, two difficulties immediately become apparent. Not enough information of the right type is available to describe adequately the state of the atmosphere in detail, and also, the mathematical equations expressing the physical laws are too complex for an exact solution unless simplifications are made.

Experience has shown that much of the atmosphere, more particularly in temperate latitudes, is often in a delicate state of dynamic equilibrium so that even quite small disturbances may be followed by rapid development and significant changes in its state. As examples, we may note the formation of thunderstorms following a localized 'trigger action' in statically unstable air, and the small beginnings of a wave on the polar front eventually developing into a major low. The original disturbance may be so small as to be missed entirely because the observing stations happen to be too widely spaced so that the development area lies undetected amongst them. This latter reason applies particularly to observations from the upper atmosphere, because radiosonde stations are normally at least 200 kilometres apart and often much more over the oceans and sparsely inhabited areas. To a considerable degree this difficulty has been obviated by the use of satellite cloud pictures (see Figure 74, Plates XIX and XX and Appendix B, Section B.2.1) although the problem of incorporating cloud pictures into a regular numerical forecasting technique has not yet been completely resolved.

12.1.2 *Finding a solution*

It is unfortunate that development should depend so critically upon details of the existing state, much of which is unobservable. This fact alone imposes a serious restriction on the feasibility of forecasting. However, considerable success in forecasting has been attained using two different approaches to the problem:

(a) the theoretical approach in which equations expressing the physical laws are solved,

(b) the empirical approach based upon the idea that, since the equations are constantly being solved naturally, observation of weather changes in the immediate past should give clues to development in the immediate future.

12.2 THEORETICAL APPROACH

12.2.1 *Numerical forecasting*

Let us consider a typical forecast problem: we want to know what the wind will be at 500 millibars over a given place 24 hours ahead. We may take the geostrophic value as being sufficiently close to the real value, and since we know this depends upon the contour pattern at 500 millibars, we must forecast this pattern 24 hours ahead. This is done by forecasting the height of the 500-millibar surface at each of a network of points lying around the original point, and then using these heights to draw the contours. Taking each point in turn, we calculate the forecast height from a knowledge of the existing height and the simplified equations describing its rate of change with time. The first 24-hour forecast cannot be done in one step without considerable loss of accuracy because the rate of change is itself not constant. Instead the calculation is split into a number of short time-steps, each sufficiently short that it can be safely assumed that little error will result from taking the rate of change as constant over each short step. Steps of 1 hour's duration are usually convenient, so that the 24 steps needed for the forecast result in a lengthy computation, each new step using as initial conditions the forecast obtained from the previous step. It is clear that, using a network which necessarily involves hundreds of points, finding a solution of the equations becomes extremely laborious. If a forecast is to be obtained in a reasonable time the calculations must be made at a high speed, and to do this electronic computers are indispensable.

Methods of forecasting similar to this are known as *numerical forecasting*; their success depends largely upon the nature of the simplifications made to the original equations. If they are too drastic, then the resulting rather theoretical atmosphere may be significantly different from the real atmosphere and the forecast changes would also be largely unreal.

Numerical forecasting was first suggested as a possibility in 1912 by V. Bjerknes, but it was not until 1922 that Richardson produced the results of the first practical attempt to use the method. This brave attempt was unpromising, and it was not until data from the upper atmosphere became more abundant after 1930, firstly from aircraft and later through the introduction of radiosondes, that further serious attacks on the problem could be made. Only then did the problem become sufficiently well understood that the necessary simplifications could be made without removing the essential factors. With the advent of electronic computers the first successful numerical forecasts were produced in 1949. The forecast consisted of a prediction of the contour pattern at a mid-tropospheric level 24 hours ahead—all further details of the forecast had to be deduced by a forecaster working in the normal manner.

Since those days a great deal of scientific effort has been put into the problem of numerical forecasting. In essence what is required is a model of the atmosphere which can be adjusted to approximate to the initial state of the real atmosphere by using actual data—coded surface observations, radiosonde ascents etc. The more complex the model, that is to say the more weather parameters it can handle, the better the approximation is likely to be, always assuming that there are sufficient data available to justify the greater complexity. We start the forecast with sets of grid values of various atmospheric parameters defining the initial state of the model atmosphere and obtain the forecast values, i.e. the grid values of these parameters at some time in the future (say 12, 24, 36, 48 . . . hours ahead) by using modifications of the mathematical equations which describe their rate of change. These forecast grid values define a future state of the model atmosphere, which is presumed to be an approximation to the future state of the real atmosphere. Again the nearer the modified equations are to the correct ones the better the forecast state is likely to be. One penalty that has to be paid for using complicated equations is that the time-step has to be much shorter—say a minute or two instead of an hour, so that the forecasting program has to be applied, say, 40 times more frequently to produce a forecast for a given period ahead. Further, with an improved model and system of equations it is possible to extend the forecast period considerably from 24 hours to, say, 72 hours ahead. In view of the normal speed of movement of weather systems, this longer period almost inevitably requires that the model must extend to the whole of the hemisphere, i.e. a large number of grid points is required. All these factors—more parameters in the model, more grid values for each parameter, more time-steps to cover a given forecast period— necessitate a large and fast computer if the forecast is to be produced in a reasonable time. Indeed some of the most powerful computers yet constructed are employed on the problem of numerical forecasting.

In general the computer output consists of grid values or sets of isopleths of such parameters as surface pressure, contour heights of pressure surfaces in the upper atmosphere (see Section 2.5.2) etc. The model used in the British Meteorological Office does produce point values of rainfall. Many of the computer products are of direct use to aviation services but the types of forecast normally issued to the general public via newspaper, television or radio are still prepared by a forecaster who naturally bases his forecast on the computer-based charts.

12.3 EMPIRICAL APPROACH

12.3.1 *Introduction*

Instead of seeking a numerical solution to the equations, we can observe how the atmosphere evolves from one state to the next and by studying a large number of synoptic situations we can obtain some clues as to the probable solution for a new situation as it arises. All we are doing is to use our experience of past weather to help us forecast new changes; we use similar processes constantly in our everyday life. In meteorology the greatest drawback to applying this method is the enormous variety of the past weather. It is hopeless to try to remember each past synoptic situation; we must help our memories by finding methods for removing unimportant detail through the use of simplified descriptions, thus making the basic ideas more readily understandable. Some of the methods which have been devised to achieve this will now be considered.

12.3.2 *Models*

Examination of sequences of charts covering long periods shows that the weather does not change in a completely random way—if it did then forecasting would be impossible. It has been found that there is a strong tendency for certain changes to follow each other in a fairly logical and orderly succession, a whole set of changes being observable on many separate occasions with only minor differences between one occasion and the next. As an example we may note the sequence of stages in the life-history of an occluding depression as described in Section 10.1. Although each depression has its own peculiarities, all have certain properties in common which can be combined into a *model*. Many models of atmospheric structures have been described including, for example, fronts, polar lows, cold pools, blocking highs and jet streams.

In olden days, before air-mass and frontal analysis were introduced, models tended to be nothing more than pressure systems, and with them attempts were made to relate the weather directly to pressure. Even our brief discussion of synoptic meteorology has been sufficient to show that there is only a poor relationship. Unfortunately, the household barometer still tends to perpetuate the idea by connecting pressure readings with the weather, described in such terms as 'change', 'wet' or 'fine'. The models introduced in recent years, although still descriptive to a large extent, do contain some ideas about the physical processes involved in their evolution. Some models have proved exceedingly useful but there has been a tendency to use them a little too slavishly, especially when the synoptic situation was unusual; they were not devised to cover all possibilities, so any tendency to overwork them should be carefully avoided.

One form of model, which on occasions has been claimed as significant, is the occurrence of *periodicities* in the weather, for example, the 'regular' occurrence of a warm summer after so many years, or of warm or cold spells at fixed dates in the year. Much effort has been devoted to the search for such periodicities but it now seems that any which may exist are rarely sufficiently regular or persistent to be used for forecasting. As obvious exceptions, however, we may note diurnal and seasonal cycles.

12.3.3 *Extrapolation*

A very useful and widely practised method of forecasting is simple *extrapolation* of past trends into the future. Careful analysis of a sequence of charts gives the velocities and accelerations of key features such as pressure centres, fronts and isobars, and their motion may then be extended into the future on the assumption that changes will continue to occur in a similar way. The greater the period of extrapolation, the greater will be the errors introduced. Up to 12 hours ahead this is probably the most accurate method of forecasting available, but it clearly cannot take account of developments which are entirely new, and since these become increasingly important as the extrapolation period increases, it is clear that this method must be used with caution.

The formulae used for determining velocities and accelerations may be either entirely empirical—rules of thumb—or based partly on physical reasoning.

12.3.4 *Analogues*

If, by looking through past records, we can find a synoptic situation similar to the one whose future changes are required, a study of the developments which followed the past situation may be used as a guide to possible changes in the present situation. The enormous variety of synoptic patterns makes it impossible to compare exactly the present with some past occasion. Even if only an approximate likeness is sought, records covering several decades are required and this in itself leads to a complex system for classifying the charts. On the whole, seeking an *analogue* from past records does not seem capable of giving good forecasts on its own, but may be useful on occasions when other methods are not satisfactory.

12.3.5 *Climatology*

The study of mean values and frequencies of meteorological quantities, such as temperature, cloud amount and rainfall, is known as *climatology*; it is concerned with the 'mean weather' and variations of the actual weather about this mean. Climatology is itself a large subject and one which is expanding too. It has many aspects including, for example, *statistical climatology*, the study of the magnitudes of the means of frequencies themselves. Also, there is *synoptic climatology*, which endeavours to explain the deviations from the mean in terms of large-scale physical processes taking place in the atmosphere. In regions like Britain where the weather is highly variable, direct use of statistical climatology in practical forecasting is of some but limited use. For example, there is little point in forecasting the maximum temperature on a July day to be 20 °C say, simply because this happens to be the mean value for July; nor is there much value in trying to improve this forecast by indicating the probable limits within which the maximum value would lie, say, 10–30 °C. However, a study of past records may be useful in providing a necessary restraint by preventing absurd forecasts. As an obvious example, a forecast of a 4 °C maximum would not be made for the July day. Other examples would be the almost certain non-occurrence of some phenomena at particular times and places, for example, snow in summer, hurricanes over Britain, or radiation fog with a gale.

In contrast to these limited applications of statistical climatology, synoptic climatology seems to be more promising especially in regard to understanding long-term trends in weather changes. It already seems that this subject is giving a useful approach to the problem of long-range forecasting—for periods of weeks or months ahead.

12.3.6 *Parameters*

The development of most weather phenomena depends upon a large number of factors or *parameters*, each of which must have a value lying within a certain range. For example, in Section 4.2.1 we saw that the formation of radiation fog depends upon certain values of wind speed, cloud amount and moisture content. Even if the physical processes involved are so complex that a complete and accurate determination of the importance of each parameter cannot be calculated theoretically, examination of past records may show which values of these parameters are important. Thus, in the example of fog formation which we considered, the exact values of the parameters which allow the fog to form vary somewhat from place to place and can only be found by studying conditions accompanying fog formation in the past. Diagrams may then be prepared showing the influence of each parameter.

This method is known as *local forecasting* since it emphasizes the effects of local features such as topography, water surfaces and smoke sources. Fog prediction is an important illustration of this method; others are thunderstorm prediction, funnelling of winds, and incidence of showers with different wind directions.

12.4 PRACTICAL FORECASTING

It is clear that the theoretical approach to the problem of forecasting is the more scientific, but there still remain obstacles to the full exploitation of the method. One of these is a lack of data from certain regions. We have seen (Section

12.2) that we need data from at least a hemisphere, if not from the whole globe, if a satisfactory forecast for a few days ahead is to be achieved. Under the auspices of the World Meteorological Organization (WMO) the main object of the World Weather Watch (WWW) is to obtain just these facilities. More radiosonde stations are planned but, as is suggested in Appendix B, we shall have to depend largely on satellites to complete the network and for world-wide communications (see Sections B.1.1 and B.2.5). It is not always a simple matter to convert a combination of satellite data and conventional observations into computer input and much scientific effort is being put into the problem of making the most effective use of the enormous amount of material being provided by the ever more sophisticated instruments carried on satellites.

The initial state of the model represented by sets of grid-point values is known as the objective analysis because it is produced by a standard computer program operating on the data available. At present the analysis may be adjusted by experienced forecasters before it is used as the starting point for the numerical forecast procedure. For example, the forecaster may use his judgment to correct or to reject an observation in the light of later data or he may invent an observation in order to ensure that the objective analysis is consistent with, say, a satellite picture.

When the senior forecaster at the Central Forecasting Office (CFO) at Bracknell comes to prepare his forecast he has at his disposal the numerical forecasts from the computer usually presented in the form of charts displaying sets of isopleths (isobars or contours), together with later observations, special observations such as radar information and satellite observations. In addition he knows whether the objective analysis has had to be based on fewer data than usual and he is aware of deficiencies in the model such as significant physical processes not incorporated or inadequately represented. From experience he has knowledge of systematic errors which turn up regularly such as weather systems being moved too quickly or too slowly. With all this extra information and expertise at his command the forecaster may modify the computer product—most frequently the 24-hour forecast surface chart on which he can put finer detail—for example the frontal analysis. When he has finalized his forecast charts, the details of the weather forecasts for Press, radio and television have to be prepared in the light of experience and the details on the latest large-scale synoptic charts.

For aviation forecasting, the forecast charts at upper levels straight from the computer have proved satisfactory for broad-scale route planning. For individual flights more detail such as the position of the jet streams may have to be added. In general, local forecasters at the many forecasting stations spread throughout the country accept the broad-scale features of the forecast presented by CFO, but from experience gained by concentrating on their own locality they can frequently add useful detail on such weather hazards as fog, frost and icy roads which vary markedly over short distances.

At present (1977) CFO prepares forecasts for three days ahead (the outlooks for the next 48 hours which follow many forecasts on radio and television are based on these) and the 48- and 72-hour numerical forecast charts are very useful though considerable modification towards the end of the period may be necessary due to known deficiencies in the model or in the initial analysis. Current numerical models have produced forecast charts for up to 7 days ahead, though little reliance can be placed on the second half of the period. Looking to the future it seems possible that this period (3–7 days ahead) can be extended to 10 or even to 14 days ahead. This will certainly require an improved and enlarged network of observational data together with very sophisticated atmospheric models in which more of the physics of the real atmosphere is represented. In turn these improvements will require larger and faster computers if the forecast is to be produced in a reasonable time. Of course, when we are considering periods of 7–14 days ahead, we can hope to forecast only the broad developments and the type of weather to be expected— we can never hope to be able to say, for example, that in 10 days' time it will start to rain in London at 9 a.m.

In the models considered above, the horizontal grid length is several hundred kilometres while in the vertical the atmosphere is represented by about 6–12 levels. Attempts are being made to use a much finer network of grid points in order to tackle the problems of local forecasting. Here there may be as many as 10–20 levels to cover the lowest 1–2 kilometres of the atmosphere with a horizontal grid length of a few tens of kilometres. This type of model can only be expected to be useful where there is a really dense network of observational data and the area covered by the model will be correspondingly small. At present (1977) this project is still in the research and development stage.

BIBLIOGRAPHY

FREEMAN, M.H.; 1970. Weather forecasting for supersonic transport. *Met Mag*, **99**, pp. 138–143.

HOWKINS, G.A.; 1973. The Meteorological Office 360/195 computing system. *Met Mag*, **102**, pp. 5–14.

KIRK, T.H.; 1974. The use of numerical forecasts. *Met Mag*, **103**, pp. 14–20.

LEE, D. and RATCLIFFE, R.A.S.; 1976. Objective methods of long-range forecasting using surface pressure anomalies. *Weather*, **31**, pp. 56–65.

MASON, B.J.; 1970. Future developments in meteorology: an outlook to the year 2000. *Q J R Met Soc*, **96**, pp. 349–368.

MASON, B.J.; 1976. Towards the understanding and prediction of climatic variations. *Q J R Met Soc*, **102**, pp. 473–498.

MURRAY, R.; 1973. Forecasting seasonal rainfall and temperature for England and Wales in spring and autumn from anomalous atmospheric circulation over the northern hemisphere. *Met Mag*, **102**, pp. 15–26.

RATCLIFFE, R.A.S.; 1973. Recent work on sea surface temperature anomalies related to long-range weather forecasting. *Weather*, **28**, pp. 106–117.

SAWYER, J.S.; 1971. Possible effects of human activity on world climate. *Weather*, **26**, pp. 251–262.

SINGLETON, F.; 1975. Human intervention in the operational objective analysis. *Met Mag*, **104**, pp. 323–330.

WICKHAM, P.G.; 1974. Some examples of the rainfall forecasts produced by the fine-mesh version of the 10-level model. *Met Mag*, **103**, pp. 209–224.

APPENDIX A

RADAR AND METEOROLOGY

A.1 INTRODUCTION

A.1.1 *Principles of radar*

In recent years radar has been developed as a powerful technique for studying the weather. Radar is an echo-sounding device in which the range of a target is measured by the time taken for a pulse of energy (in the form of electro-magnetic waves) to travel from a transmitter to a target and back to the receiver. Usually the same aerial is used for transmission as for reception. The radiation is transmitted in pulses, as the outgoing beam must be switched off to obtain detectable echoes in the incoming beam. For satisfactory resolution the beam must be narrow (about 1–2 degrees) and so the source of radiation is placed at the focus of a paraboloid aerial. The power of the reflected signal depends on many factors such as the power of the transmitted pulse, the pulse length, the aerial size, the range and nature of the target, attenuation of the signal and so on. In practice the choice of radar has to be made as a compromise taking into account cost and the use to be made of the radar. For example the power of the back-scattered energy and the degree of atmospheric attenuation of the radiation travelling to and from the target both increase with decrease of wavelength. For detection of precipitation a wavelength in the range 30 to 200 millimetres is chosen. Again the power of the return signal is proportional to the pulse length, whereas the resolution in range is inversely proportional to it; so again a compromise is necessary (a pulse length of 5 microseconds gives a range resolution of 750 metres). The resolution in the cross-range direction is proportional to the beam width which is itself directly proportional to the wavelength and inversely proportional to the linear dimensions of the aerial. Hence for a required resolution, the longer the wavelength the larger the aerial. The cost of an aerial increases rapidly with its size and the shorter the wavelength the greater the atmospheric attenuation, so again a compromise has to be reached.

A.2 METEOROLOGICAL RADAR

A.2.1 *Radar techniques*

There are five main types of radar in use and these will be discussed briefly.

(a) *Tracking radar.* This is the type of radar used to measure upper winds. A reflector is attached to a balloon filled with hydrogen or helium and as the balloon is released and rises, the aerial (often called the antenna, or dish) is directed on the azimuth and elevation of the reflector and the range of the reflector recorded. From readings on two consecutive positions the average height and speed of the balloon and hence the velocity of the air at that average height can be determined.

(b) *Qualitative radar.* This type of radar is the one commonly associated with weather radar—it is the one available to the forecasters at London Weather

Centre, for example—and displays a qualitative picture of precipitation. There are two common ways of displaying on a radar screen the echo pattern from clouds, or more correctly from the precipitation particles within clouds. The particles constituting most of the cloud are too small to produce a measurable echo.

(1) *Plan position indicator* (PPI). This gives the position of echoes in an almost horizontal plane surrounding the radar set. The radar beam scans continuously about a vertical axis and points in a direction just a few degrees above the horizon so that echoes are received from clouds at progressively higher levels as the distance from the set increases and it is possible that the precipitation particles observed at relatively high levels may not be reaching the ground. The normal maximum working range is about 150 kilometres. Another form of display is obtained by using Constant Altitude PPI (CAPPI). The aerial goes through a cycle of continuous rotation in azimuth with a step up in elevation after each rotation. By using several displays, one for each altitude, a set of constant altitude maps can be produced and a three-dimensional picture of precipitation obtained.

(2) *Range height indicator* (RHI). Here the radar dish is set on a fixed bearing and nods in a vertical plane, thus producing a cross-section through the atmosphere. From it the heights of bases and tops of echoes can be found.

Although a wavelength of 30 millimetres has been used for weather radar because a relatively small (and therefore cheap) aerial will provide the required resolution, this wavelength does lead to attenuation problems in that moderate or heavy rainfall close to the radar will blot out any echoes from a greater range on the same azimuth. A wavelength of 100 millimetres is now widely used and is an excellent wavelength for the detection of precipitation. Aerials to give a beam width of 1° are not prohibitively large and attenuation problems are virtually non-existent. The disadvantage is that equipment is bulky and expensive. In the last few years a wavelength of 60 millimetres has been favoured for airborne radar where the restriction of aerial size requires a wavelength shorter than 100 millimetres.

In radar circles the various wavelength bands are known as K band (10 mm), X band (30 mm), C band (50 mm), S band (100 mm) and L band (200 mm).

(c) *Quantitative radar.* This is essentially a PPI weather radar carefully sited to measure rates of precipitation and hence, by time integration, total precipitation over an area up to the size of a catchment area. The vertical beam width must be small and the wavelength sufficiently long to avoid serious attenuation problems. This method of precipitation measurement has the advantages that:

(1) Measurements over an area are available at one point in real time (and most users, if not all, require measurements over an area).

(2) Snowfall can be measured as accurately as rainfall.

(3) Short-term forecasts of precipitation movement and intensity can be made.

Disadvantages are the cost and maintenance of the equipment itself together with some uncertainty over the translation of the radar records into amounts of precipitation at the ground.

To check the feasibility of this method a weather radar of 100-millimetre wavelength was set up in North Wales and four authorities, the Meteorological Office, the Water Resources Board, the Dee and Clwyd River Authority and Plessey Radar Limited have been collaborating in the investigation—the Dee Weather Radar Project. Much useful information has been obtained and it appears that, particularly if the radar records are checked by one or more rain-gauges strategically placed, the method might be viable on an operational basis.

(d) *Ultra-sensitive radar*. The ordinary weather radar will detect precipitation particles, but not the majority of cloud particles. At the other extreme, with a very sensitive radar, that is one with a powerful transmitter, a sensitive receiver and a large aerial, it is possible to detect echoes from a clear sky. Some of these come from birds and insects, but the majority come from irregular refractive index gradients. Small-scale variations in the refractive index of the air depend mainly on variations in temperature and humidity: each has about the same importance at low levels, but temperature changes dominate in the dry air aloft.

With such a radar it is possible to detect convection both before and after the development of cumulus clouds and also to study gravity waves, lee waves and clear-air turbulence.

(e) *Doppler radar*. If a radar target is moving towards or away from the radar, the frequency of the signal received after reflection is, because of the Doppler effect, slightly different from the transmitted frequency. Doppler radar is designed to interpret this effect in terms of the radial velocity of the target. As an illustration we will consider here the use of Doppler techniques where the targets are precipitation particles. When the radar beam is pointed vertically, the mean Doppler velocity in a pulse volume (i.e. the volume defined by the beam width and the pulse length) is the combined result of the vertical velocity of the air and the mean fall-speed of the targets relative to the air. If the drop-size distribution is assumed or measured in the rain as it hits the ground a mean fall-speed can be calculated and the vertical air velocity deduced. This method may be used to assess vertical motion in convective cells, but in widespread vertical motion the velocity is too small and is better derived from the divergence of the horizontal wind, measured by using Doppler radar in the conical scanning mode. In this mode, which is particularly useful in regions where the airflow is horizontally almost homogeneous over distances of about 10 kilometres, the beam is rotated in azimuth with the elevation angle kept constant throughout a complete rotation. The line-of-sight velocity of targets is measured simultaneously within a number of range positions as the beam describes circular paths at different altitudes. If this velocity is plotted against azimuth for a complete rotation of a single range position a velocity–azimuth display (VAD) is obtained. From an analysis of the VAD we can deduce the horizontal wind and the divergence of the horizontal wind (see Section 2.6.3). These values will represent average values over the area of the scanned circle. By integrating the divergence in the vertical it is possible to calculate the vertical air velocity. In performing these calculations we have to allow for the target (precipitation particles) fall-speeds which will of course contribute to the radial velocities measured round the scanned circles. On many occasions it is sufficient to measure fall-speeds at the centre of the circles between successive conical scans by using the radar beam pointing vertically and to assume that these values are applicable round the scanned circles. In addition the component of radial velocity due to fall-speeds can be minimized by keeping the elevation angle of the radar beam as small as possible.

A.2.2 *Problems studied by radar*

In listing the various types of radar, we have to some extent mentioned the sorts of problems that can be tackled with the help of a meteorological radar. It is, however, worth stressing how intractable some of these would be if radar were not available. For example consider the problems faced by a forecaster who is providing short-range forecasts (say 3–6 hours ahead) for an area within 50–100 kilometres of his station. His main tool will be his set of synoptic charts (see Section 9.1.2) and he will be lucky if he has available hourly reports from more than half a dozen stations within his area. In a showery situation, much of the showery activity may well miss the relatively few reporting stations and the forecaster will have to prepare his forecast from general principles rather than from detailed knowledge. With a weather radar providing PPI displays he can monitor the position and intensity of precipitation areas as frequently as he chooses and observe their movement and development. To obtain such a detailed picture from a set of observing stations would require so close a network as to be prohibitive on the grounds of cost alone.

This problem and also the determination of the vertical velocity of the air within a few tens of kilometres of the radar set are both mesoscale problems, that is they lie somewhere between the scale of a single cumulus or cumulonimbus cloud and the scale of the lows and highs on a synoptic chart and for this scale of event meteorological radar is a most powerful tool.

BIBLIOGRAPHY

BROWNING, K.A.; 1972. Atmospheric research using the Defford radar facility. *Weather*, **27**, pp. 2–13.

HARROLD, T.W.; 1966. The measurement of rainfall using radar. *Weather*, **21**, pp. 247–249 and 256–258.

HARROLD, T.W. and NICHOLASS, C.A.; 1972. The accuracy of some recent radar estimates of surface precipitation. *Met Mag*, **101**, pp. 193–205.

HARROLD, T.W., ENGLISH, E.J. and NICHOLASS, C.A.; 1973. The Dee weather radar project: the measurement of area precipitation using radar. *Weather*, **28**, pp. 332–338.

HARROLD, T.W., ENGLISH, E.J. and NICHOLASS, C.A.; 1974. The accuracy of radar-derived rainfall measurements in hilly terrain. *Q J R Met Soc*, **100**, pp. 331–350.

SIMPSON, J.E.; 1967. Aerial and radar observations of some sea-breeze fronts. *Weather*, **22**, pp. 306–316.

TAYLOR, B.C. and BROWNING, K.A.; 1974. Towards an automated weather radar network. *Weather*, **29**, pp. 202–216.

APPENDIX B

WEATHER SATELLITES

B.1 INTRODUCTION

B.1.1 *General*

Satellites are already playing an important role in observing the weather on a global scale and this role will have to be extended if we are to take full advantage of more powerful computers and more complex mathematical models of the atmosphere. With these tools it seems likely that reliable forecasts of the main weather patterns for up to a week ahead will be possible provided adequate observations covering at least a hemisphere, if not the whole globe, are available. At present the radiosonde network providing data in the upper air is adequate only over North America and Europe and it is clear that, over the oceans at least, we shall have to depend on satellites to complete this network on a global scale. In Section B.2 below some of the ways in which satellites can provide observational data are listed; the list is by no means complete. Satellite technology is constantly changing and becoming more sophisticated so that a continued improvement in quality and quantity of data can be expected.

B.1.2 *Brief historical review*

The first satellite (Sputnik) was launched by the USSR in 1957 and since that date the USA and the USSR have each launched more than a score of weather satellites, that is satellites carrying scientific instruments designed to measure and transmit to the ground various geophysical parameters of the earth–atmosphere system. At first rocket power was small and rockets had to be launched so that the rotation of the earth would give extra impetus to lift them above the earth's atmosphere. Clearly the maximum effect is on a launching at the equator in an easterly direction. However, a satellite in such an orbit can obtain measurements from equatorial regions only.

The initial series of American satellites was called by the name TIROS (*Television and Infra-Red Observation Satellite*). TIROS I was launched in April 1960 into an orbit inclined at about 48° to the equator providing useful meteorological data between about 55°N and 55°S. By 1965 near polar orbits were achieved and soon after it was possible to make polar orbits sun-synchronous so that a satellite passes over the same place on the earth's surface at the same time each day. In fact the satellite passes over each place twice a day, once in daylight and once in darkness. While the TIROS series orbit at a height of about 800 kilometres, for later series of American satellites the orbits were raised to about 1400 kilometres. The period of rotation is rather under 2 hours.

The Russian satellites, both the experimental series (COSMOS) and the operational series (METEOR) starting in 1969 have orbits mainly in the height range 600 to 900 kilometres.

On the earliest satellites the camera pointed in a fixed direction so that as the satellite completed an orbit, the camera was sometimes pointing towards the earth's surface and sometimes away from it; but since the launching of TIROS X in 1965, cameras have been mounted in such a way that pictures are taken when the camera is pointing vertically downwards. At the start of the TIROS series information was transmitted from the satellite to the ground only when the satellite was within range of two stations in North America. Because of limited storage capacity, information from a limited area only was available and this needed quite lengthy processing before it was of practical use to the meteorologist. As a result there was a delay of some hours between the time when information was recorded and the time of its reception in Britain. This delay coupled with the limited area made the information of very limited operational use to the forecasters in Britain. However in 1964 American satellites were successfully provided with the *automatic picture transmission* (APT) facility. This is a device whereby any ground station with the necessary receiving and reproduction equipment can receive these pictures in real time and make them almost immediately available to the forecaster.

While a polar orbiting satellite can survey the whole of the earth's surface sequentially, there are advantages in also having satellites which can survey a particular part of the earth's surface continually, i.e. satellites which are stationary with respect to the earth—geostationary satellites. Such a satellite rotating with the earth must be in the equatorial plane at a height of about 36 000 kilometres. The United States *Advanced Technology Satellites* (ATS) launched in 1966 and 1967 into such an orbit are still operating. Although not specifically meteorological satellites, they each carry a camera which views a wide strip of the earth and can provide about every 30 minutes a picture of the cloud systems covering nearly a quarter of the globe. ATS I is located at 150°W over the central Pacific and ATS III has been positioned at several longitudes between 45°W over the Atlantic and 85°W over the Amazon basin. The first geosynchronous meteorological satellite SMS 1 (*Synchronous Meteorological Satellite*) was launched in May 1974. Although high latitudes are out of the field of view of these satellites, a set of five geostationary satellites equally spaced above the equator can maintain continual surveillance over approximately 75 per cent of the earth's surface.

In May 1967, UK3 (ARIEL III) was launched by an American rocket from California. This was the first completely British satellite, although two earlier American satellites, ARIEL I in 1962 and ARIEL II in 1964 had carried British instruments. UK3 carried five experiments sponsored by different groups of scientists. The Meteorological Office experiment concerned the vertical distribution of molecular oxygen density in the upper atmosphere. The satellite gave two weeks of useful data; this was in fact its expected life.

B.2 USES OF METEOROLOGICAL SATELLITES

B.2.1 *Cloud photography*

This was the first source of meteorological information transmitted from satellites starting with TIROS I and still remains probably the most important satellite aid in day-to-day weather forecasting. When a cloud picture is received an essential first step is to superpose on it a grid of latitude and longitude; this is a simple problem when the path of the orbit and the exact time of the picture are known. In Britain we can receive pictures from three successive passes of a polar orbiting satellite. The first pictures are from a pass over eastern Europe, the second, some 2 hours later, from an overhead pass and the third, after a further 2 hours, from a pass over the Atlantic. These pictures can be combined into a composite cloud picture (nephanalysis) covering most of Europe and the Mediterranean area and extending westwards to cover most of the northern North Atlantic. On many occasions the analysis of the current synoptic chart has been improved by a study of such a nephanalysis. In the absence of cloud, areas of sea ice and land areas covered by ice and snow can be discerned on the photographs (see Plates XIX and XX).

In the current American ITOS (*Improved TIROS Operational Satellite*) satellites the APT system has been replaced by scanning radiometers which produce both visible and infrared images so that cloud pictures can be made available at night as well as during the day. The airborne radiometer scans from horizon to horizon across the track of the satellite although only the central 1000-kilometre section is reliable. With this system the receiving station picks up a continuous picture as long as the satellite is within range. The ground resolution at the sub-satellite point is 4 kilometres in the visible and 8 kilometres in the infrared. The ITOS IV satellite launched in 1972 carries in addition a very high resolution scanning radiometer producing a high quality picture with a resolution of about 0.9 kilometre in both visible and infrared channels.

B.2.2 *Measurement of temperature*

The ITOS IV launched into a near polar orbit and the ATS VI launched in 1974 into a geosynchronous orbit each carry very high resolution radiometers which can be used to estimate the temperature of cloud tops or that of the underlying surface in the absence of cloud by measuring the infrared radiation emitted in the 11 μm atmospheric window. Difficulties arise in deducing the temperature owing to absorption by water vapour and carbon dioxide (even in the atmospheric window) and broken cloud is a problem when estimating the surface temperature. Nevertheless surface temperature can be determined to within about 1–2 °C. If a vertical temperature profile is measured or assumed, then the temperature of the cloud top will provide an estimate of its height.

B.2.3 *Measurement of vertical temperature profiles*

Within the spectrum of terrestrial radiation there are ranges of wavelengths which are absorbed by certain constituent gases of the atmosphere, principally carbon dioxide, ozone and water vapour. Carbon dioxide absorbs and radiates strongly in a band centred on 15 μm and is mixed uniformly through the atmosphere up to a height of about 80 kilometres. Basically the absorption of radiation by a gas is dependent upon the pressure and temperature of the gas and the wavelength of the radiation. From a study of the absorption due to carbon dioxide at various heights a vertical profile of temperatures can be constructed. If an instrument on a satellite measures radiation at a specific wavelength information on the absorption at a particular pressure and hence at a particular height can be obtained and the temperature of the gas deduced. By using a number of narrow wavebands near 15 μm a vertical profile of temperature can be built up. Because we know the vertical distribution of carbon dioxide in the atmosphere, we can use radiation measurements to deduce the temperature profile. If we know the vertical temperature profile we can use such measurements to deduce the vertical distribution of a non-uniformly mixed absorber, for example water vapour.

In the SIRS (*Satellite Infra-Red Spectrometer*) experiments eight different wavelengths are used including one at which there is no absorption by carbon dioxide; this provides information on the temperature of the cloud top or of the underlying surface in the absence of clouds. A crude temperature profile can be constructed between about 24 kilometres and the cloud top. This is representative of an area about 200 kilometres square. With this technique a radiometer has a poor vertical resolving power, that is, a deduced temperature is more closely related to a layer of the atmosphere than to an exact height. For this reason SIRS data have been widely used in estimating the thickness of a layer of the atmosphere between standard isobaric surfaces, more particularly the 1000–500-millibar thickness (see Section 2.5.4) even when cloud is present and the lowest part of the temperature profile has had to be extrapolated. In view of the sparseness of upper-air data, especially over the southern hemisphere and the oceans generally, this is a very important step forward and considerable effort is going into improving this system.

The latest series of American satellites has carried a *Vertical Temperature Profile Radiometer* (VTPR). This is an 8-channel scanning radiometer with an improved horizontal resolution of 60×60 kilometres. By assuming the horizontal uniformity of the atmosphere and surface over a much larger area it is possible to obtain a temperature profile and hence a thickness down to the ground when broken cloud is present. This technique can be used for sea areas only, but of course this is where the largest gaps in the radiosonde network occur.

B.2.4 *Measurement of upper winds*

Measurement of upper winds from satellites is most important in the tropics where air flow cannot be predicted from the temperature (thickness) field (see Section 2.5.2) and sequences of cloud pictures from a geostationary satellite have been used to estimate the wind. In principle the method is perfectly straightforward. The satellite camera is stationary relative to the earth so that in a sequence of pictures the displacements of a recognizable cloud target will provide a measure of the cloud velocity and hence of the wind speed. In practice difficulties arise in choosing a suitable cloud target and in assigning a height to the wind deduced. The target must persist throughout the sequence, preferably without too much growth or decay, and the movement of the cloud must reflect the movement of the air so that, for example, lee-wave clouds must be avoided. At present it is assumed that low-level clouds move with the wind at the level of the cloud base which is assigned as 900 millibars. For higher clouds the height is inferred from infrared measurements giving the cloud temperature.

Outside the tropics winds have been measured in the southern hemisphere by means of tracking balloons. These are inextensible superpressurized balloons which drift with the wind along surfaces of constant atmospheric density and respond to interrogation from a satellite. The satellite measures the Doppler shift of the signal and by comparison of position on successive orbits the displacement of the balloon and hence the mean wind speed can be calculated. Loss of balloons by icing at low levels limits the choice of height to one above about 200 millibars (12 kilometres) if an economic 3-month life of the balloon is to be obtained. At this level a small percentage of the balloons flown lasted more than a year.

Although useful wind data are obtained with either of these techniques, neither meets the requirement for a vertical wind profile providing winds at several levels at approximately the same time and place. It is hoped to obtain such profiles shortly by the mother-balloon dropsonde system. In this system a balloon floating at about 30 millibars carries a number of sondes and a geostationary satellite relays commands to the balloon and data from balloon to the ground. On command a sonde is released, falls on a parachute and relays Omega navigation signals to the balloon and hence to the satellite and the ground. These low-frequency navigation signals from two pairs of ground stations enable the position of the sonde to be fixed very accurately. On the assumption that the parachute and hence the sonde moves horizontally with the velocity of the air, a wind profile can be constructed from a series of fixes spread over about 30 minutes.

B.2.5 *Data collection*

As more and more observational data become available, so does the problem of collection and dissemination grow. Satellites are becoming ever more necessary for the location and interrogation of instrumented balloons, ocean buoys and unmanned land stations in remote areas and indeed for world-wide communication generally.

BIBLIOGRAPHY

BARRETT, E.C.; 1967. Weather satellite cloud photography of the British region. *Weather*, **22**, pp. 151–159.

CARRUTHERS, G.P., MAY, B.R., MILLER, D.E. and STEWART, K.H.; 1973. Some current work on meteorological satellites. *Met Mag*, **102**, pp. 258–265.

HOLDSWORTH, N.; 1970. Report on a seminar on weather satellite cloud photographs, High Wycombe, October 1969. *Met Mag*, **99**, pp. 86–90.

MASON, B.J.; 1974. The contribution of satellites to the exploration of the global atmosphere and to the improvement of weather forecasting. *Met Mag*, **103**, pp. 181–201.

MOREL, P. and BANDEEN, W.; 1973. The Eole experiment: early results and current objectives. *Bull Amer Met Soc*, **54**, pp. 298–306.

POTHECARY, I.J.W. and RATCLIFFE, R.A.S.; 1966. Satellite pictures of an old occluded depression and their usefulness in analysis and forecasting. *Met Mag*, **95**, pp. 332–339.

RAO, P.K. and McLAIN, E.P.; 1974. Images from the NOAA-3 very high resolution radiometer over the North Sea and adjoining countries. *Weather*, **29**, pp. 436–442.

WORLD METEOROLOGICAL ORGANIZATION; 1973. The use of satellite pictures in weather analysis and forecasting. *WMO Tech Notes* No. 124.

BOOKS FOR FURTHER READING

BARRY, R.G. and CHORLEY, R.J.; 1971. Atmosphere, weather and climate, 2nd edn. London, Methuen and Co. Ltd.

CARTER, C.; 1971. The blizzard of '91. Newton Abbot, David and Charles.

DOBSON, G.M.B.; 1968. Exploring the atmosphere, 2nd edn. Oxford, The Clarendon Press.

DUNN, G.E. and MILLER, B.I.; 1964. Atlantic hurricanes. Louisiana, State University Press.

FLOHN, H.; 1973. Climate and weather. London, Weidenfeld and Nicolson.

FORSDYKE, A.G.; 1970. The weather guide. London, Hamlyn.

GEIGER, R.; 1965. The climate near the ground. Cambridge, Mass., Harvard University Press.

INWARDS, R.; 1950. Weather lore. London, Rider and Company.

LAMB, H.H.; 1964. The English climate. London, English Universities Press.

LANE, F.W.; 1966. The elements rage. Newton Abbot, David and Charles.

McINTOSH, D.H. and THOM, A.S.; 1969. Essentials of meteorology. London, Wykeham Publications.

MASON, B.J.; 1975. Clouds, rain and rainmaking, 2nd edn. London, Cambridge University Press.

MASSACHUSSETTS INSTITUTE OF TECHNOLOGY; 1971. Inadvertent climate modification. Report of the Study of Man's Impact on Climate. London, MIT Press.

MAUNDER, W.J.; 1970. The value of the weather. London, Methuen and Co. Ltd.

METEOROLOGICAL OFFICE; 1969. Observer's Handbook, 3rd edn. London, HMSO.

METEOROLOGICAL OFFICE; 1972. The meteorological glossary, 5th edn. London, HMSO.

MINNAERT, M.; 1954. The nature of light and colour in the open air. New York, Dover Publications, Inc.

PALMÉN, E. and NEWTON, C.W.; 1969. Atmospheric circulation systems. International Geophysics Series, 13, New York, Academic Press.

SCHONLAND, B.F.J.; 1950. The flight of thunderbolts. Oxford, Clarendon Press.

SCORER, R.S.; 1972. Clouds of the world. Newton Abbot, David and Charles.

INDEX